Higher Combinatorics

NATO ADVANCED STUDY INSTITUTES SERIES

Proceedings of the Advanced Study Institute Programme, which aims at the dissemination of advanced knowledge and the formation of contacts among scientists from different countries

The series is published by an international board of publishers in conjunction with NATO Scientific Affairs Division

A	Life Sciences	Plenum Publishing Corporation
B	Physics	London and New York
C	Mathematical and Physical Sciences	D. Reidel Publishing Company Dordrecht and Boston
D	Behavioral and Social Sciences	Sijthoff International Publishing Company Leiden
E	Applied Sciences	Noordhoff International Publishing Leiden

Series C - Mathematical and Physical Sciences

Volume 31 – Higher Combinatorics

Higher Combinatorics

Proceedings of the NATO Advanced Study Institute held in Berlin (West Germany), September 1-10, 1976

edited by

MARTIN AIGNER

Freie Universität Berlin, Fachbereich Mathematik,
Königin-Luise-Strasse 24/26, D-1000 Berlin 33, B.R.D.

D. Reidel Publishing Company

Dordrecht-Holland/Boston-U.S.A.

Published in cooperation with NATO Scientific Affairs Division

ISBN-13: 978-94-010-1222-5 e-ISBN-13: 978-94-010-1220-1
DOI: 10.1007/978-94-010-1220-1

Published by D. Reidel Publishing Company
P.O. Box 17, Dordrecht, Holland

Sold and distributed in the U.S.A., Canada, and Mexico
by D. Reidel Publishing Company, Inc.
Lincoln Building, 160 Old Derby Street, Hingham, Mass. 02043, U.S.A.

TABLE OF CONTENTS

PREFACE

It is general consensus that Combinatorics has developed into a full-fledged mathematical discipline whose beginnings as a charming pastime have long since been left behind and whose great significance for other branches of both pure and applied mathematics is only beginning to be realized. The last ten years have witnessed a tremendous outburst of activity both in relatively new fields such as Coding Theory and the Theory of Matroids as well as in more time honored endeavors such as Generating Functions and the Inversion Calculus. Although the number of text books on these subjects is slowly increasing, there is also a great need for up-to-date surveys of the main lines of research designed to aid the beginner and serve as a reference for the expert.

It was the aim of the Advanced Study Institute "Higher Combinatorics" in Berlin, 1976, to help fulfill this need. There were five sections: I. Counting Theory, II. Combinatorial Set Theory and Order Theory, III. Matroids, IV. Designs and V. Groups and Coding Theory, with three principal lecturers in each section. Expanded versions of most lectures form the contents of this book. The Institute was designed to offer, especially to young researchers, a comprehensive picture of the most interesting developments currently under way. It is hoped that these proceedings will serve the same purpose for a wider audience.

Sincere thanks are due to the Freie Universität Berlin and, in particular, to the North Atlantic Treaty Organization for their generous support which made this Institute possible. It is our hope that after Nijenrode 1974 and Berlin 1976 there will be another Advanced Study Institute on Combinatorics in the near future.

Berlin, October 1976 Martin Aigner

LIST OF PARTICIPANTS

Aigner, M., II. Math. Inst., Freie Universität, Königin-Luise-Str. 24-26, 1000 Berlin 33
Andrews, G.E., 410 McAllister Bldg., University Park, Pa. 16802, U.S.A.
Arditti, J.C., 25 rue Damesme, 75013, Paris 13, France
Assmus, E.F., 238 Uncas St., Bethlehem, Pa. 18015, U.S.A.
Baclawski, K., Dept. of Math., MIT, Cambridge, Mass. 02139, U.S.A.
Bandelt,H.-J., FB IV, Universität Oldenburg, Ammerländer Heerstr., D-2900 Oldenburg, Germany
Beth, T., Inst. f. Informatorik I, Bismarckstr. 6, 8520 Erlangen, W. Germany
Beutelsbacher, A., Math. Inst. d. Universität, Saarstr. 21, 6500 Mainz, W. Germany
Björner, A., Kräpvägen 24, 18131 Lidingö, Sweden
Böckmann, D., II. Math. Inst., Freie Universität, Königin-Luise-Str. 24-26, 1000 Berlin 33
Bogart, K. P., Math. Dept., Dartmouth College, Hanover, N.H. 03755, U.S.A.
Brylawski, T., Dept. of Math., Univ. of North Carolina, Chapel Hill, N.C. 27514, U.S.A.
Burmeister, P., FB 4, AG 1, Techn. Hochschule, Schlossgartenstr. 7, 6100 Darmstadt, W. Germany
Cameron, P.J., Merton College, Oxford OX1 4JD, U.K.
Camion, P., 3, rue F. Couperin, 78370 Plaisir, France
Cerasoli, M., Istituto di Matematica-Universita-67100 L'Aquila, Italy
Cetnicelik, A., Nuclear Energy Forum of Turkey, P.O.Box 37-Bankanhklar Ankara, Turkey
de Clerck, F., Seminar of Geometry, Univ. of Ghent, Krygslaan 271, 9000 Ghent, Belgium
Cohen, D.J.A., Dept. Math., Northeastern Univ., Boston, Mass. 02115, U.S.A.
Columbic, M., Courant Inst., New York University, 251 Mercer St., New York 10012, U.S.A.
Day, A., Dept. of Math., McMasters Univ., Hamilton, Ont., Canada
Debroey, I., Dept. W.N.F., Limburgs Universitaire Centrum, Universitaire Campus, 3610 Diepenbeek, Belgium

Dehon, M., Dept. de Math., Université Libre de Bruxelles, Blvd. de Triomphe, 1050 Brussels, Belgium
Désarménien, J., Dept. de Math., Univ. de Strasbourg, 7 rue René Descartes, 6784 Strasbourg, France
Deuber, W., Fakultät f. Mathematik, Universität, 4800 Bielefeld, W. Germany
Deza, M., 3, rue Duzas, 75008 Paris, France
Dowling, T.A., Dept. Math., Ohio State Univ., Columbus, Ohio 01234, U.S.A.
Doven, J., Dept. of Math., Univ. of Brussels, Campus Plaine CP 216, Blvd. de Triomphe, B-1050, Brussels, Belgium
Dürre, R., Technische Univ. München, Arcisstr. 21, D-8 München-2, W. Germany
Duke, R., Math. Faculty, Open Univ., Milton Keynes, U.K.
Durdel, W., II. Math. Inst., Freie Universität Berlin, Königin-Luise-Str. 24-26, 1000 Berlin 33, W. Germany
Ecker, A., Hahn-Meitner-Inst., Glienickerstr. 100, 1000 Berlin 39, W. Germany
Eichhorn, K. von, Hofmannstr. 17, 8520 Erlangen
Erdös, P., Univ. of Calgary, Math. Dept., Calgary, Alberta, Canada
Ewald, H.G., II. Math. Inst., Freie Universität, Königin-Luise-Str. 24-26, 1000 Berlin 33, W. Germany
Faigle, U., Dept. of Math., Univ. of North Carolina, Chapel Hill, N.C. 27514, U.S.A.
Foata, D., Dept. de Math., Univ. de Strasbourg, 67084 Strasbourg, France
Ganter, B., TH Darmstadt, FB4 AG1, Hochschulestr. 1, 6100 Darmstadt, W. Germany
Gardiner, A., Univ. of Birmingham, Pure Matib, Birmingham B15ZTT, U.K.
Goethals, J.M., MBLE Res. Lab., Ave. van Becelaere 2, 1170 Brussels, Belgium
Grammer, F., Rennbühlweg 59, 8431 Wolfstein, W. Germany
Greene, C., 135,S.19th Street, Apt. 1204 Philadelphia, Pa. 19103, U.S.A.
Grüning, K., Nachtweide 8, 6085 Nauheim, W. Germany
Guy, R., Dept. Math., Univ. of Calgary, Calgary (44), Alberta, Canada
Harzheim, E., Pallenbergstr. 23, 5000 Köln 60, W. Germany
Hauptmann, W., Math. Inst. Ciessen, Arndtstr. 2, 6300 Giessen, W. Germany
Heden, O., Univ. of Stockholm, Dept. of Math., Box 6701, 11385 Stockholm, Sweden
Heise, W., Math. Inst., Technische Univ. München, Arcisstr. 21, D-8 München-2, W. Germany
Higman, D.G., Dept. of Math., Univ. of Michigan, Ann. Arbor, Mich. 48104, U.S.A.
Hobbs, B.F., Point Loma College, 3900 Lomaland Drive, San Diego, Calif. 92106, U.S.A.
Høholdt, T., Math. Inst., Techn. Univ. of Denmark, 2800 Lyngby, Denmark

Holladay, K.W., Cal. Tech. Pasadena, 253-37 Slaon Lab. 91125, U.S.A.
Ingleton, A.W., Balliol College, Oxford OX1, 3BJ, U.K.
Jungnickel, D., II. Math. Inst., Freie Universität, Königin-Luise-Str. 24-26, 1000 Berlin 33, W. Germany
Kapur, K.C., Wayne State Univ., 640 Putnam, Dept. of Indust. Engineering, Detroit, Mich. 48202, U.S.A.
Kerber, A., Lehrstuhl f. Math., RWTH Aachen, Templergraben 55, 5100 Aachen, W. Germany
Kestenbrand, B., New York Inst. of Technology, Math. Dept., Old Westbury, N.Y. 11568, U.S.A.
Knapp, W., Math. Inst. d. Univ., Auf der Morgenstelle 10, 7400 Tübingen, W. Germany
König, R.,
Kovacs, P., FU Berlin, Belzigerstr. 34, 1000 Berlin 62, W. Germany
Las Vergnas, M., 19 bis rue Louis Blanc, 92250 La Garenne-Colombes, France
Lautemann, C., FU Berlin, Belzigerstr. 34, 1000 Berlin 62, W. Germany
Leeb, K., Inst. f. Informatorik I, Bismarckstr. 6, 8520 Erlangen W. Germany
Leemans, H., State Univ. at Ghent, Seminar of Geometry, Krygslaan 271, 59 9000 Ghent, Belgium
Lefèvre, Dept. of Math., CP 216, Free Univ. of Brussels, Campus-Plaine, 1050 Brussels, Belgium
Lenz, H., II. Math. Inst., Freie Universität, Königin-Luise-Str. 24-26, 1000 Berlin 33, W. Germany
Liebler, R.A., Math., Colorado State Univ., Fort Collins, Colo. 80523, U.S.A.
Lindström, B., Dept. of Math., Emaljvägen 14 V., 17573 Järfälla, Sweden
Lint, J.H. van, T.H. Eindhoven, P.O. Box 513, Eindhoven, The Netherlands
Mader, W., II. Math. Inst., Freie Universität, Königin-Luise-Str. 24-26, 1000 Berlin 33
Mason, J.H., Open University, Milton Keynes, U.K.
Matthews, L.R., Math. Inst. Oxford Univ., 24-29, St. Giles, Oxford, OX13LB, U.K.
Meissner, J., II. Math. Inst., Freie Universität, Königin-Luise-Str. 24-26, 1000 Berlin 33, W. Germany
Metz, R., FB4, Math. TU Darmstadt, Schlossgartenstr. 7, 6100 Darmstadt, W. Germany
Mielants, W., Seminar of Algebra, Univ. of Ghent, Krygslaan 276, 9000 Ghent, Belgium
Neumaier, A., Seestr. 117, 1000 Berlin 65, W. Germany
Nitz, M., II. Math. Inst., Freie Universität, Königin-Luise-Str. 24-26, 1000 Berlin 33, W. Germany
Nowakowski, R., Dept. of Math., Univ. of Calgary, Calgary, Alberta, Canada
Oxley, J.G., Math. Inst. Univ. of Oxford, 24-29 St. Giles, Oxford OX1 3LB, U.K.
Pabst, D., Schleissheimerstr. 219, 8000 München 40, W. Germany

Paustian, H., Mommsenstr. 64, 1000 Berlin 12, W. Germany
Peele, R., Univ. of North Carolina, Chapel Hill, U.S.A.
Percsy, F., Univ. de l'Etat à Mons, Ave. Maistrian 15, 7000 Mons, Belgium
Pfundt, D., Philipp-Reiss-Str. 21, 8520 Erlangen, W. Germany
Podewski, K.-P., Inst. f. Math. d. TU Hannover, Welfengarten 1, 3000 Hannover, W. Germany
Poguntke, W., Fachbereich Math.-Techn. Hochschule, 6100 Darmstadt, W. Germany
Prins, G., Dept. of Math., Wayne State Univ., Detroit, Mich. 48202, U.S.A.
Rado, R., Math. Dept. of the University, Reading, U.K.
Randow, R. von, Inst. f. Ökonometrie + Operations, Research d. Univ. Bonn, Postfach 589, 5300 Bonn 1, W. Germany
Razen, R.A., Inst. f. Math., Montanuniv.Leoben, 8700 Leoben, Austria
Rosumek, R., II. Math. Inst., Freie Universität, Königin-Luise-Str. 24-26, 1000 Berlin 33, W. Germany
Roehmel, J., II. Math. Inst., Freie Universität, Königin-Luise-Str. 24-26, 1000 Berlin 33, W. Germany
Schäfer, W., Werner-v.-Siemens-Str. 19, 8520 Erlangen, W. Germany
Scheid, H., Gesamtschule Wuppertal, Dietr.-Bonhoeffer-Weg 1, 5600 Wuppertal, W. Germany
Schöne, J., II. Math. Inst., Freie Universität, Königin-Luise-Str. 24-26, 1000 Berlin 33, W. Germany
Schulz, R.-H., II. Math. Inst., Freie Universität, Königin-Luise-Str. 24-26, 1000 Berlin 33, W. Germany
Schweigert, D., Math. Inst., Universität Kaiserslautern/Trier
Spencer, J., Dept. of Math., MIT, Cambridge, Mass. 02139, U.S.A.
Stanley, R., Math. Dept., MIT, Cambridge, Mass. 02139, U.S.A.
Steffens, K., Inst. f. Math., TU Hannover, Weltengarten 1, 3000 Hannover, W. Germany
Strehl, V., Informatorik I., Univ., Erlangen, Bismarckstr. 6, 8520 Erlangen, W. Germany
Thas, J.A., Seminar of Higher Geometry, Univ. of Ghent, Krygslaan 271, 9000 Ghent, Belgium
Thiele, E.-J., Math. Inst. Technische Univ. Berlin, Strasse des 17. Juni 135, 1000 Berlin 12
Thomas, G., FU Berlin, Fachbereich Theoret. Medizin, Wiss. Einrichtung, Hindenburgdamm 30, 1000 Berlin
Verheiden, E., California Inst. of Technology, CIT 253-37, Pasadena, Calif. 91125, U.S.A.
de Vries, G., Lehrstuhl f. Numer. u. Angewandte Math., Lotzestr. 16-18, 3400 Göttingen, W. Germany
Wagner, B., Lehrstuhl D. f. Math., RWTH Aachen, Wilhemstr. 45, 5100 Aachen, W. Germany
Weiss, R., II. Math. Inst., Freie Universität, Königin-Luise-Str. 24-26, 1000 Berlin 33, W. Germany
White, N.L. Math. Dept., Univ. of Florida, Gainesville, Fla. 32611, U.S.A.

Wille, R., AG 1, Fachbereich Math., Techn. Hochschule, 6100 Darmstadt, W. Germany
Wille, D., Inst. f. Math./TU Hannover, Weltengarten 1, 3000 Hannover, W. Germany
Wilson, R., The Ohio State Univ., Columbus, Ohio 43210, U.S.A.
de Winne, P., Seminare Hogere Meetkunde, Rijksuniversiteit Gent, Krygslaan 271, 9000 Gent, Belgium
Wolff, K., Math. Inst. Giessen, Arndtstr. 2, 6300 Giessen, W. Germany
Wolfmann, J., Centre Universitaire de Toulon, 83130 La Garde, France
Wolfsdorf, K., TU Berlin, Fachbereich Mathematik

PART I

COUNTING THEORY

ENUMERATION OF PARTITIONS: THE ROLE OF EULERIAN SERIES AND q-ORTHOGONAL POLYNOMIALS

George E. Andrews[1] and Richard Askey[2]

Department of Mathematics, Pennsylvania State University, University Park, Pennsylvania 16802
and
Department of Mathematics, University of Wisconsin, Madison, Wisconsin 53706

ABSTRACT. The theory of partitions has long been associated with so called basic hypergeometric functions or Eulerian series. We begin with discussion of some of the lesser known identities of L.J. Rogers which have interesting interpretations in the theory of partitions. Illustrations are given for the numerous ways partition studies lead to Eulerian series. The main portion of our work is primarily an introduction to recent work on orthogonal polynomials defined by basic hypergeometric series and to the applications that can be made of these results to the theory of partitions. Perhaps it is most interesting to note that we deduce the Rogers-Ramanujan identities from our solution to the connection coefficient problem for the little q-Jacobi polynomials.

1. SOME THEOREMS OF L.J. ROGERS

L.J. Rogers [20] became famous for the following two identities that were proved by him in 1894.

---

(1) Partially supported by the United States Army under Contract No. DAAG29-75-C-0024 and National Science Foundation Grant MSP74-07282.
(2) Partially supported by the United States Army under Contract No. DAAG29-75-C-0024 and National Science Foundation Grant MCS75-06687-3.

M. Aigner (ed.), Higher Combinatorics, 3-26.

$$1 + \sum_{n=1}^{\infty} \frac{q^{n^2}}{(1-q)(1-q^2)\dots(1-q^n)} = \prod_{n=0}^{\infty} \frac{1}{(1-q^{5n+1})(1-q^{5n+4})} \tag{1.1}$$

$$1 + \sum_{n=1}^{\infty} \frac{q^{n^2+n}}{(1-q)(1-q^2)\dots(1-q^n)} = \prod_{n=0}^{\infty} \frac{1}{(1-q^{5n+2})(1-q^{5n+3})} \tag{1.2}$$

The story of the neglect of these identities and their subsequent rediscovery by S. Ramanujan has been told many times [17; pp.90-91], [8; Ch.7], so we shall not dwell upon it here. Our current object is to examine several other results of Rogers that are still little known. These identities are nearly as striking as the "Rogers-Ramanujan identities" (i.e. (1.1) and (1.2)) and their partition-theoretic counterparts are even more surprising. These results and the general discussion of such problems in Section 2 lead us to the conclusion that an extensive investigation of the analytic theory of Eulerian series is long overdue. In Section 3, we make some contributions to such a study by illustrating how orthogonal polynomials defined by basic hypergeometric functions are useful in the study of partitions; the results in this Section are merely an introduction to an extensive study of such orthogonal polynomials in [9].

1.1 The analytic form of Rogers's identities

Rogers (in [20] and [21]) gave numerous series-product identities. To avoid dilution of the interest of these results by exhibiting too many, we choose six which well illustrate these amazing discoveries of Rogers:

$$1 + \sum_{n=1}^{\infty} \frac{q^{n^2}}{(1-q)(1-q^2)\dots(1-q^{2n})}$$

$$= \prod_{n=0}^{\infty} \frac{1}{(1-q^{2n+1})(1-q^{20n+4})(1-q^{20n+16})}, \tag{1.1.1}$$

$$1 + \sum_{n=1}^{\infty} \frac{q^{n^2+2n}}{(1-q)(1-q^2)\dots(1-q^{2n+1})}$$

$$= \prod_{n=0}^{\infty} \frac{1}{(1-q^{2n+1})(1-q^{20n+8})(1-q^{20n+12})}, \tag{1.1.2}$$

$$1 + \sum_{n=1}^{\infty} \frac{q^{n^2+n}}{(1-q)(1-q^2)\ldots(1-q^{2n})}$$

$$= \prod_{\substack{n=1 \\ n\not\equiv \pm 1,\pm 8,\pm 9,10 \pmod{20}}}^{\infty} (1-q^n)^{-1}, \tag{1.1.3}$$

$$1 + \sum_{n=1}^{\infty} \frac{q^{n^2+n}}{(1-q)(1-q^2)\ldots(1-q^{2n+1})}$$

$$= \prod_{\substack{n=1 \\ n\not\equiv \pm 3,\pm 4,\pm 7,10 \pmod{20}}}^{\infty} (1-q^n)^{-1}; \tag{1.1.4}$$

$$\{\prod_{m=1}^{\infty} (1+q^{2m})\}\{1 + \sum_{n=1}^{\infty} \frac{q^{n^2}}{(1-q^4)(1-q^8)\ldots(1-q^{4n})}\}$$

$$= \prod_{n=0}^{\infty} \frac{1}{(1-q^{5n+1})(1-q^{5n+4})}; \tag{1.1.5}$$

$$\{\prod_{m=1}^{\infty} (1+q^{2m})\}\{1 + \sum_{n=1}^{\infty} \frac{q^{n^2+2n}}{(1-q^4)(1-q^8)\ldots(1-q^{4n})}\}$$

$$= \prod_{n=0}^{\infty} \frac{1}{(1-q^{5n+2})(1-q^{5n+3})}. \tag{1.1.6}$$

For reference we note that (1.1) and (1.2) first appeared in [20; p.328, eq.(1) and p.329, eq.(2)]. Equation (1.1.1) is the second equation on page 331 of Roger's memoir [20], while equation (1.1.2) is equivalent to equation (7) on page 331 of [20]. Equation (1.1.3) is equivalent to the equation appearing just after equation (12) on page 332 of [20], and equation (1.1.4) is equivalent to equation (6) on page 331 of [20]. Equation (1.1.5) is the last equation on page 330 of [20], and equation (1.1.6) is the identity preceding equation (7) on page 331 of [20].

1.2 The partition-theoretic implications of Rogers's identities.

Section 2 will discuss the actual techniques used to shift

from combinatorial to analytic identities. Here we confine ourselves to an examination of the quite unexpected partition theorems that have arisen from (1.1.1)-(1.1.6).

We begin by recalling

Theorem 1. (Euler's theorem). The number of partitions $b_1+b_2+\ldots+b_r$ of n where $b_1 \geq b_2 \geqq b_3 \geqq \ldots$ and each b_i is odd equals the number of partitions $c_1+c_2+\ldots+c_s$ of n where $c_1 > c_2 > c_3 > \ldots$.

A standard exercise in elementary partition theory (cf. [3; Ch.13],[18; Ch.19]) shows that Euler's theorem is easily derived from the trivial infinite product identity

$$\prod_{n=1}^{\infty} (1+q^n) = \prod_{n=1}^{\infty} \frac{(1-q^{2n})}{(1-q^n)} = \prod_{n=1}^{\infty} \frac{1}{1-q^{2n-1}} .$$

B. Gordon [16] observed that (1.1.1) may be interpreted as follows:

Theorem 2. (Gordon's theorem). The number of partitions $b_1+b_2+\ldots+b_r$ of n where $b_1 \geqq b_2 \geqq b_3 \geqq \ldots$ and each b_i is odd or $\equiv \pm 4 \pmod{20}$ equals the number of partitions $c_1+c_2+\ldots+c_3$ of n where $c_1 > c_2 \geqq c_3 > c_4 \geqq c_5 > \ldots$.

Thus Theorem 2 seems quite closely related to Theorem 1; however its proof lies much deeper. There is a similar interpretation of (1.1.2) mentioned by Gordon [16] and explicitly given by W. Connor [13]:

Theorem 3. The number of partitions $b_1+b_2+\ldots+b_r$ of n where $b_1 \geqq b_2 \geqq b_3 \geqq \ldots$ and each b_i is odd or $\equiv \pm 8 \pmod{20}$ equals the number of partitions $c_1+c_2+\ldots+c_{2s+1}$ of n into an odd number of parts where $c_1 \geqq c_2 \geqq c_3 > c_4 \geqq c_5 > c_6 \geqq c_7 > \ldots$ (i.e. $c_i \geqq c_{i+1}$ with strict inequality if i is odd and $\geqq 3$).

W. Connor [13] has also interpreted (1.1.3) and (1.1.4):

Theorem 4. The number of partitions $b_1+b_2+\ldots+b_r$ of n where $b_1 \geqq b_2 \geqq b_3 \geqq \ldots$ and each b_i is $\not\equiv \pm 1, \pm 8, \pm 9, 10 \pmod{20}$ equals the number of partitions $c_1+c_2+\ldots+c_s$ of n into an even number of parts where $c_1 \geqq c_2 > c_3 \geqq c_4 > c_5 \geqq \ldots$.

Theorem 5. The number of partitions $b_1+b_2+\ldots+b_r$ of n where $b_1 \geqq b_2 \geqq b_3 \geqq \ldots$ and each $b_i \not\equiv \pm 3, \pm 4, \pm 7, 10 \pmod{20}$ equals the number of partitions $c_1+c_2+\ldots+c_s$ of n where $c_1 \geqq c_2 > c_3 \geqq c_4 > c_5 \geqq c_6 > \ldots$.

Equations (1.1.5) and (1.1.6) have apparently not been interpreted before; we note that the infinite product in (1.1.5) is the same one appearing in (1.1), and that in (1.1.6) is the same as in (1.2):

Theorem 6. The number of partitions of n with distinct parts and with each even part larger than twice the number of odd parts equals the number of partitions of n into parts congruent to 1 or 4 modulo 5.

Theorem 7. The number of partitions of n into distinct parts each larger than 1 in which each even part is larger than twice the number of odd parts equals the number of partitions of n into parts congruent to 2 or 3 modulo 5.

As is well-known, equation (1.1) implies that the number of partitions of n into parts that differ by at least 2 also equals the number of partitions of n into parts congruent to 1 or 4 modulo 5.

Example. Let us consider the case $n = 14$ in Theorem 6 and the first Rogers-Ramanujan identity:

distinct parts, each even > twice no. odd parts	parts differ by at least 2	parts ≡ 1,4 (mod 5)
14	14	14
13+1	13+1	11+1+1+1
12+2	12+2	9+4+1
11+3	11+3	9+1+1+1+1+1
10+4	10+4	6+6+1+1
10+3+1	10+3+1	6+4+4
9+5	9+5	6+4+1+1+1+1
8+6	9+3+1	6+1+...+1
8+5+1	8+6	4+4+4+1+1
8+4+2	8+5+1	4+4+1+...+1
7+6+1	8+4+2	4+1+...+1
6+5+3	7+5+2	1+1+...+1.

Since the Rogers-Ramanujan identities have so far resisted a purely combinatorial proof, we must certainly expect that a combinatorical proof of Theorem 6 would be difficult; however the following problem is conceivably more tractable:

Problem 1. Provide a bijection between the partitions of n in which the parts differ by at least 2 and the partitions of n into distinct parts in which each even part is larger than twice the number of odd parts.

2. EULERIAN SERIES GENERATING FUNCTIONS

The connection between Theorems 2-7 and identities (1.1.1)-(1.1.6) is easily made. We choose the passage from (1.1.1) to Theorem 2 as an example. Recall that

$$\frac{1}{(1-q)(1-q^2)\dots(1-q^{2n})} \tag{2.1}$$

is the generating function for partitions in which there are at most 2n parts (see Theorems 12-2 and 13-1 of [3]). Furthermore

$$q^{n^2} = q^{0+1+1+2+2+\dots+(n-1)+(n-1)+n}. \tag{2.2}$$

Thus if $\gamma_1+\gamma_2+\dots+\gamma_j$ is an arbitrary partition subject to $j \leqq 2n$, we may define $\gamma_{j+1} = \gamma_{j+2} = \dots = \gamma_{2n+1} = 0$ and we see that $\gamma_1+n = c_1$, $\gamma_2+(n-1) = c_2$, $\gamma_3+(n-1) = c_3,\dots,\gamma_{2n-1}+1 = c_{2n-1}, \gamma_{2n}+1 = c_{2n}$, $\gamma_{2n+1} = c_{2n+1}$ yields an arbitrary partition $c_1+c_2+c_3+\dots$ that has either $2n$ or $2n+1$ nonzero parts. Thus the left side of (1.1.1) is clearly the generating function for partitions $c_1+c_2+c_3+\dots$ subject to $c_1 > c_2 \geqq c_3 > c_4 \geqq \dots$. Since the right side of (1.1.1) clearly generates partitions in which each part is odd or $\equiv \pm 4 \pmod{20}$, we see that the identity of the generating functions in (1.1.1) implies the truth of Theorem 2.

The arguments used to establish Theorems 3-5 are quite similar; the simple summation identities that replace (2.2) are respectively

$$q^{n^2+2n} = q^{1+1+2+2+\dots+n+n+n};$$
$$q^{n^2+n} = q^{1+1+2+2+\dots+n+n};$$
$$q^{n^2+n} = q^{0+1+1+2+2+\dots+n+n}.$$

To deduce Theorems 6 and 7 from (1.1.5) and (1.1.6) we have to be a little more careful. We illustrate with (1.1.5)

$$\prod_{m=1}^{\infty} (1+q^{2m}) \sum_{n=0}^{\infty} \frac{q^{n^2}}{(1-q^4)(1-q^8)\dots(1-q^{4n})}$$

$$= \prod_{m=1}^{\infty} (1+q^{2m}) \sum_{n=0}^{\infty} \frac{q^{n^2}}{(1-q^2)(1-q^4)\dots(1-q^{2n})(1+q^2)(1+q^4)\dots(1+q^{2n})}$$

$$= \sum_{n=0}^{\infty} \frac{q^{n^2}}{(1-q^2)(1-q^4)\ldots(1-q^{2n})} \cdot \prod_{m=1}^{\infty} (1+q^{2n+2m}). \tag{2.3}$$

In the same way we treated the left side of (1.1.1), we see that

$$\frac{q^{n^2}}{(1-q^2)(1-q^4)\ldots(1-q^{2n})} = \frac{q^{1+3+5+\ldots+2n-1}}{(1-q^2)(1-q^4)\ldots(1-q^{2n})}$$

is the generating function for partitions with n parts each odd and distinct from the rest.

Since $\prod_{m=1}^{\infty} (1+q^{2n+2m})$ generates the partitions into distinct even parts where each part is $> 2n$, we see that

$$\frac{q^{n^2}}{(1-q^2)(1-q^4)\ldots(1-q^{2n})} \prod_{m=1}^{\infty} (1+q^{2n+2m}) \tag{2.4}$$

generates partitions with distinct parts wherein n parts are odd and each even part is $> 2n$. Summing expressions (2.4) for all $n \geqq 0$ yields the appropriate generating function (2.3).

Finally we note that Theorem 6 may now be deduced from (1.1.5) and the fact that the right side of (1.1.5) generates partitions with parts $\equiv 1,4 \pmod 5$.

Theorem 7 follows from (1.1.6) in the same manner; one must now use the identity $3+5+\ldots+(2n+1) = n^2+2n$.

2.1 Some open problems on generating functions

Our seven theorems give the impression that partition identities are proved by first simply obtaining appropriate generating functions and then applying analytic techniques to prove the identity of the required generating functions.

In this short subsection we shall mention some open problems concerning the "simple" passage from the partition-theoretic identity to the analytic identity.

First we mention the "linked" partition ideals that have been studied extensively elsewhere ([5],[7]; a complete introduction to them will appear as Chapter 8 in [8]). In summary, linked partition ideals give rise to classes c of partitions whose generating functions satisfy linear homogeneous q-difference equations with polynomial coefficients: This means that if $f_c(x;q) = \Sigma P_c(m,n)x^m q^n$ where $P_c(m,n)$ is the number of partitions of n with m parts that are in the "linked partition ideal c", then there exists N and polynomials $\pi_i(x,q)$ such that

$$\sum_{j=0}^{\infty} f_c(xq^j;q)\pi_j(x;q) = 0. \tag{2.1.1}$$

This state of affairs leads us to the following problem [5; p.1037]:

"What finite linear q-difference equations with polynomial coefficients (such as (2.1.1)) have solutions that can be represented by "higher dimensional" q-series?"

In general it appears to be very difficult to answer this question. However even very specific problems (for which we know the answer to the above question), still leave us with mysteries. For example, let $b_{k,a}(m,n)$ denote the number of partitions of n into m parts: $b_1+b_2+\ldots+b_m$, with $b_i \geq b_{i+1}$, $b_i-b_{i+k-1} \geq 2$ and $b_{m-a+1} > 1$; let $B_{k,a}(n) = \sum_{m\geq 0} b_{k,a}(m,n)$ so that $B_{k,a}(n)$ is the total number of such partitions without regard for the number of parts.

B. Gordon [15] (see also [1]) has proved:

Theorem 8. For $1 \leq a \leq k$ and all n, $A_{k,a}(n) = B_{k,a}(n)$ where $A_{k,a}(n)$ is the number of partitions of n into parts $\not\equiv 0, a \pmod{2k+1}$.

We remark that the cases $k = a = 2$ and $k = a+1 = 2$ are the standard partition-theoretic forms of the Rogers-Ramanujan identities (1.1) and (1.2). We also note that there is a corresponding analytic identity [6]:

$$\sum_{n_1,n_2,\ldots,n_{k-1}\geq 0} \frac{q^{N_1^2+N_2^2+\ldots+N_{k-1}^2+N_a+N_{a+1}+\ldots+N_{k-1}}}{(q)_{n_1}(q)_{n_2}\ldots(q)_{n_{k-1}}}$$

$$= \prod_{\substack{n=1 \\ n\not\equiv 0,\pm a \pmod{2k+1}}}^{\infty} (1-q^n)^{-1}, \tag{2.1.1}$$

where $(a)_n = (a;q)_n = (1-a)(1-aq)\ldots(1-aq^{n-1})$, and $N_j = n_j+n_{j+1}+\ldots+n_{k-1}$; furthermore this reduces to (1.1) when $k = a = 2$ and to (1.2) when $k = a+1 = 2$.

Finally we point out that one may indirectly prove that

$$\sum_{m,n\geq 0} b_{k,a}(m,n)x^m q^n$$

$$= \sum_{n_1,n_2,\ldots,n_{k-1}\geq 0} \frac{x^{N_1+N_2+\ldots+N_{k-1}} q^{N_1^2+N_2^2+\ldots+N_{k-1}^2+N_a+\ldots+N_{a-1}}}{(q)_{n_1}(q)_{n_2}\cdots(q)_{n_{k-1}}}. \tag{2.1.2}$$

Thus (2.1.2) constitutes a "higher-dimensional" q-series representation of the generating function related to one set of partitions considered in Theorem 8.

When $k = 2$, we may easily prove (2.1.2) in the direct manner used to produce the generating functions for Theorems 2-7. When $k > 2$, the only known proof of (2.1.2) consists of showing that each side is the unique solution of a certain system of q-difference equations.

<u>Does there exist a direct combinatorial proof of</u> (2.1.2)?

3. q-ORTHOGONAL POLYNOMIALS AND PARTITIONS

In Section 2 we illustrated the relationship between q-series and partition generating functions. Now we propose to prove the eight identities (1.1),(1.2),(1.1.1)-(1.1.6) as corollaries of some new results of ours on orthogonal polynomials defined by basic hypergeometric series. In order to keep our survey to a reasonable length we shall consider only one family of q-orthogonal polynomials, the little q-Jacobi polynomials. Previously the adjective "little" has not been used; however, we have found a second q-analog of the Jacobi polynomials which has a third free parameter so we call these latter polynomials the big q-Jacobi polynomials.

$$p_n(x;\alpha,\beta|q) = {}_a\phi_1\left(\begin{matrix} q^{-n},\alpha\beta q^{n+1};q,qx \\ \alpha q \end{matrix}\right), \tag{3.1}$$

where

$${}_{r+1}\phi_r\left(\begin{matrix} a_1,a_2,\ldots,a_{r+1};q,t \\ b_1,b_2,\ldots,b_r \end{matrix}\right)$$

$$= \sum_{n\geq 0} \frac{(a_1)_n (a_2)_n \cdots (a_{r+1})_n t^n}{(q)_n (b_1)_n \cdots (b_r)_n}, \tag{3.2}$$

and

$$(A)_n = (1-A)(1-Aq)\ldots(1-Aq^{n-1}). \tag{3.3}$$

The zeroth little q-Jacobi polynomial is 1; the next two are

$$p_1(x;\alpha,\beta|q) = 1 - \frac{(1-\alpha\beta q^2)}{(1-\alpha q)} x, \tag{3.4}$$

$$p_2(x;\alpha,\beta|q)$$

$$= 1 - \frac{(1-\alpha\beta q^3)}{(1-\alpha q)} (1+q)xq^{-1} + \frac{(1-\alpha\beta q^3)(1-\alpha\beta q^4)}{(1-\alpha q)(1-\alpha q^2)} x^2 q^{-1}. \tag{3.5}$$

The primary results that we shall assume from the theory of basic hypergeometric functions are the q-analog of the binomial series [11; p.66, eq.(4)]:

$$\sum_{n=0} \frac{(a)_n z^n}{(q)_n} = \frac{(az)_\infty}{(z)_\infty},$$

and the summation formula for the balanced ${}_3\phi_2$ ([11; p.68,eq.(1)]; the series in (3.2) is called balanced [10; p.56] if one of the a_i is q^{-n} where n is a nonnegative integer and $qa_1a_2\ldots a_{r+1} = b_1b_2\ldots b_r$):

$${}_3\phi_2\left(\begin{matrix} q^{-n}, Aq^n, B \\ C, \frac{ABq}{C} \end{matrix}; q,q\right) = \frac{B^n (\frac{Aq}{C})_n (\frac{C}{B})_n}{(C)_n (\frac{ABq}{C})_n}. \tag{3.7}$$

There are numerous theorems that one can prove about these polynomials, and we shall present an extensive account in [9]. Here we are only concerned with obtaining the connection coefficient formula (Theorem 10). To this end we must prove an intermediate result (Theorem 9) which fully describes the orthogonality relationship for the little q-Jacobi polynomials. The existence of this result is shown by W. Hahn [18; §10]; W. Al-Salam and A. Verma [22] prove this result from a consideration of the moments of the related distribution.

Theorem 9.

$$\sum_{i=0}^{\infty} \frac{\alpha^i q^i (q^{i+1})_\infty}{(\beta q^{i+1})_\infty} p_n(q^i;\alpha,\beta|q) p_m(q^i;\alpha,\beta|q)$$

$$= \begin{cases} 0 & \text{if} \quad m \neq n \\ \dfrac{\alpha^n q^n (q)_\infty (\alpha\beta q^{n+1})_\infty (q)_n}{(\beta q^{n+1})_\infty (\alpha q)_\infty (\alpha q)_n (1-\alpha\beta q^{2n+1})}, & m = n. \end{cases} \tag{3.8}$$

Proof. Since $p_m(x;\alpha,\beta|q)$ is of degree m in x, we see that we may replace $p_m(q^i;\alpha,\beta|q)$ by any other polynomial in q^i of degree m that we choose; after treating this simplified case it is easy to deduce (3.8).

$$\sum_{i=0}^{\infty} \frac{\alpha^i q^i (q^{i+1})_\infty}{(\beta q^{i+1})_\infty} p_n(q^i;\alpha,\beta|q) q^{im}$$

$$= \sum_{i=0}^{\infty} \frac{\alpha^i q^{(m+1)i} (q^{i+1})_\infty}{(\beta q^{i+1})_\infty} \sum_{j=0}^{n} \frac{(q^{-n})_j (\alpha\beta q^{n+1})_j q^{(i+1)j}}{(q)_j (\alpha q)_j}$$

$$= \frac{(q)_\infty}{(\beta q)_\infty} \sum_{j=0}^{n} \frac{(q^{-n})_j (\alpha\beta q^{n+1})_j q^j}{(q)_j (\alpha q)_j} \sum_{i=0}^{\infty} \frac{(\beta q)_i}{(q)_i} (\alpha q^{m+j+1})^i$$

$$= \frac{(q)_\infty}{(\beta q)_\infty} \sum_{j=0}^{n} \frac{(q^{-n})_j (\alpha\beta q^{n+1})_j q^j}{(q)_j (\alpha q)_j} \frac{(\alpha\beta q^{m+j+2})_\infty}{(\alpha q^{m+j+1})_\infty} \qquad \text{(by (3.6))}$$

$$= \frac{(q)_\infty (\alpha\beta q^{m+2})_\infty}{(\beta q)_\infty (\alpha q^{m+1})_\infty} \sum_{j=0}^{n} \frac{(q^{-n})_j (\alpha\beta q^{n+1})_j (\alpha q^{m+1})_j q^j}{(q)_j (\alpha q)_j (\alpha\beta q^{m+2})_j}$$

$$= \frac{(q)_\infty (\alpha\beta q^{m+2})_\infty (\beta^{-1} q^{-n})_n (q^{-m})_n}{(\beta q)_\infty (\alpha q^{m+1})_\infty (\alpha q)_n (\alpha^{-1}\beta^{-1} q^{-n-m-1})_n} \qquad \text{(by (3.7))}$$

$$= \frac{(q)_\infty (\alpha\beta q^{m+n+2})_\infty (q^{-m})_n \alpha^n q^{n(m+1)}}{(\beta q^{n+1})_\infty (q^{m+1})_\infty (\alpha q)_n}$$

$$= \begin{cases} 0 & \text{if} \quad 0 \leqq m < n \\ \dfrac{(-\alpha)^n q^{n(n+1)/2} (q)_\infty (\alpha\beta q^{2n+2})_\infty (q)_n}{(\beta q^{n+1})_\infty (\alpha q)_\infty}, & \text{if} \quad m = n. \end{cases} \tag{3.9}$$

Hence the top line of (3.8) is valid. Futhermore since the leading coefficient of $p_n(x;\alpha,\beta|q)$ is $(-1)^n(\alpha\beta q^{n+1})_n q^{-n(n-1)/2}/(\alpha q)_n$, we may multiply the bottom line of (3.9) by this expression to obtain the bottom line of (3.8). Thus Theorem 9 is established.

We remark that in our proof of the connection formula, equation (3.9) is as important as (3.8).

There are numerous reasons for considering the problem of connection coefficients. Szegö [24; Ch.9] used the connection coefficients between Jacobi polynomials with one altered parameter to find the lowest order Cesaro mean which sums every Jacobi series of continuous functions. D. Feldheim seems to have been the first to obtain an explicit formula (a ${}_3F_2$ hypergeometric function) for the connection coefficients between two arbitrary Jacobi polynomials (see [10; Ch.7] for an extensive account of connection coefficient problems). Theorem 10 is a q-analog of Feldheim's theorem.

Theorem 10. If $a_{k,n}$ is defined for all n with $0 \leqq k \leqq n$ by

$$p_n(x;\alpha,\beta|q) = \sum_{k=0}^{n} a_{k,n} p_k(x;\alpha,\beta|q), \tag{3.10}$$

then

$$a_{k,n} = \frac{(-1)^k q^{k(k+1)/2} (\gamma\delta q^{n+1})_k (q^{-n})_k (\alpha q)_k}{(q)_k (\gamma q)_k (\alpha\beta q^{k+1})_k}$$

$$\cdot\ {}_3\phi_2\left(\begin{matrix} q^{-n+k}, \gamma\delta q^{n+k+1}, \alpha q^{k+1} & ;q,q \\ \gamma q^{k+1}, \alpha\beta q^{2k+2} & \end{matrix}\right). \tag{3.11}$$

Proof. The orthogonality relationship in Theorem 9 allows us to immediately find a representation for $a_{k,n}$:

$$a_{k,n} = \frac{\sum_{i=0}^{\infty} \frac{\alpha^i q^i (q^{i+1})_\infty}{(\beta q^{i+1})_\infty} p_k(q^i,\alpha,\beta|q)p_n(q^i;\gamma,\delta|q)}{\sum_{i=0}^{\infty} \frac{\alpha^i q^i (q^{i+1})_\infty}{(\beta q^{i+1})_\infty} (p_k(q^i;\alpha,\beta|q))^2}, \tag{3.12}$$

and the denominator of the right side of (3.12) is immediate from (3.8) with $m = n = k$. We now evaluate the numerator expression in (3.12):

$$\sum_{i=0}^{\infty} \frac{\alpha^i q^i (q^{i+1})_\infty}{(\beta q^{i+1})_\infty} p_k(q^i;\alpha,\beta|q)p_n(q^i;\gamma,\delta|q)$$

$$= \sum_{j=0}^{n} \frac{(q^{-n})_j(\gamma\delta q^{n+1})_j q^j}{(q)_j\ (\gamma q)_j} \sum_{i=0}^{\infty} \frac{\alpha^i q^i (q^{i+1})_\infty}{(\beta q^{i+1})_\infty} p_k(q^i;\alpha,\beta|q)q^{ij}$$

$$= (q)_\infty \sum_{j=0}^{n} \frac{(q^{-n})_j(\gamma\delta q^{n+1})_j q^j}{(q)_j\ (\gamma q)_j} \frac{(\alpha\beta q^{j+k+2})_\infty (q^{-j})_k \alpha^k q^{k(j+1)}}{(\beta q^{k+1})_\infty (\alpha q^{j+1})_\infty (\alpha q)_k}$$

(by the penultimate line in the equation string (3.7))

$$= \frac{(-1)^k (q)_\infty \alpha^k q^{k(k+1)/2}}{(\beta q^{k+1})_\infty (\alpha q)_k} \sum_{j=0}^{n-k} \frac{(q^{-n})_{j+k}(\gamma\delta q^{n+1})_{j+k} q^{j+k}(\alpha\beta q^{j+2k+2})_\infty}{(q)_j\ \ (\gamma q)_{j+k}\ \ (\alpha q^{j+k+1})_\infty}$$

$$= \frac{(-1)^k (q)_\infty (\gamma\delta q^{n+1})_k (\alpha\beta q^{2k+2})_\infty (q^{-n})_k \alpha^k q^{k(k+3)/2}}{(\beta q^{k+1})_\infty (\alpha q)_\infty (\gamma q)_k}$$

$$ {}_3\phi_2\left(\begin{matrix} q^{-n+k}, \gamma\delta q^{n+k+1}, \alpha q^{k+1}; q, q \\ \gamma q^{k+1}, \alpha\beta q^{2k+2} \end{matrix}\right). \tag{3.13}$$

To obtain $a_{k,n}$ we divide the last expression in the string of equations (3.13) by the bottom line of (3.8) with $m = n = k$. Hence

$$a_{k,n} = \frac{(-1)^k q^{k(k+1)/2}(\gamma\delta q^{n+1})_k (q^{-n})_k (\alpha q)_k}{(\alpha\beta q^{k+1})_k (q)_k (\gamma q)_k}$$

$$_3\phi_2\left(\begin{matrix} q^{-n+k}, \gamma\delta q^{n+k+1}, \alpha q^{k+1} & ;q,q \\ \gamma q^{k+1}, \alpha\beta q^{2k+2} & \end{matrix}\right), \tag{3.14}$$

which is the desired result.

Theorem 10 may be immediately generalized to yield an identity for the functions $_{r+1}\phi_r$ defined in (3.2).

Theorem 11. With $a_{k,n}$ as given by (3.9),

$$_{r+2}\phi_{r+1}\left(\begin{matrix} q^{-n}, \gamma\delta q^{n+1}, a_1, \ldots, a_r; q, xq \\ \gamma q, \quad b_1, \ldots, b_r \end{matrix}\right)$$

$$= \sum_{k=0}^{n} a_{k,n}\ {}_{r+2}\phi_{r+1}\left(\begin{matrix} a^{-k}, \alpha\beta q^{k+1}, a_1, \ldots, a_r; q, xq \\ \alpha q, \quad b_1, \ldots, b_r \end{matrix}\right).$$

Proof. We simply observe that since (3.10) is a polynomial identity in x it remains valid if we replace x^k throughout by any other expression in particular

$$\frac{(a_1)_k \cdots (a_r)_k x^k}{(b_1)_k \cdots (b_r)_k}.$$

We shall now illustrate the power of Theorem 11 by deducing from it, Watson's q-analog of Whipple's theorem [25]:

Theorem 12.

$$ {}_8\phi_7\left(\begin{matrix} a,q\sqrt{a},-q\sqrt{a},b,c,d,e,q^{-n} & ;q,\frac{a^2q^{n+2}}{bcde} \\ \sqrt{a},-\sqrt{q},\frac{aq}{b},\frac{aq}{c},\frac{aq}{d},\frac{aq}{e},aq^{n+1} & \end{matrix}\right) $$

$$ = \frac{(aq)_n(\frac{aq}{de})_n}{(\frac{aq}{d})_n(\frac{aq}{e})_n}\,{}_4\phi_3\left(\begin{matrix} \frac{aq}{bc},d,e,q^{-n};q,q \\ ,\frac{de}{aq^n},\frac{aq}{b},\frac{aq}{c} \end{matrix}\right). $$

Proof. In (3.15) we take $r = 2$, $\beta = \delta$, $a_1 = \alpha q$, $x = 1$, $b_2 = q^2\alpha\delta a_2/b_1$. We motivate these choices by noting that these substitutions reduce both $a_{k,n}$ and the ${}_4\phi_3$ on the right side of (3.15) to balanced ${}_3\phi_2$-series which are summable by (3.7). Thus (3.15) is reduced to

$$ {}_4\phi_3\left(\begin{matrix} q^{-n},\gamma\delta q^{n+1},\alpha q,a_2 ;q,q \\ \gamma q,b_1,\frac{q^2\alpha\delta a_2}{b_1} \end{matrix}\right) $$

$$ = \sum_{k=0}^{n} \frac{(-1)^k q^{k(k+1)/2}(\gamma\delta q^{n+1})_k(q^{-n})_k(\alpha q)_k}{(\alpha\delta q^{k+1})_k(q)_k(\gamma q)_k} $$

$$ \cdot\ {}_3\phi_2\left(\begin{matrix} q^{-n+k},\gamma\delta q^{n+k+1},\alpha q^{k+1} & ;q,q \\ \gamma q^{k+1} \quad ,\alpha\delta q^{2k+2} & \end{matrix}\right) $$

$$ \cdot\ {}_3\phi_2\left(\begin{matrix} q^{-k},\alpha\delta q^{k+1},a_2 ;q,q \\ b_1,\frac{q^2\alpha\delta a_2}{b_1} \end{matrix}\right) $$

$$ = \sum_{k=0}^{n} \frac{(-1)^k q^{k(k+1)/2}(\gamma\delta q^{n+1})_k(q^{-n})_k(\alpha q)_k}{(\alpha\delta q^{k+1})_k(q)_k(\gamma q)_k} $$

$$ \cdot\frac{(\alpha q^{k+1})^{n-k}\quad(\delta q^{k+1})_{n-k}(\frac{\gamma}{\alpha})_{n-k}}{(\gamma q^{k+1})_{n-k}(\alpha\delta q^{2k+2})_{n-k}}\cdot\frac{a_2^k(\frac{\alpha\delta q^2}{b_1})_k(\frac{b_1}{a_2})_k}{(b_1)_k(\frac{\alpha\delta a_2 q^2}{b_1})_k}. $$

$$= \frac{\alpha^n q^{n^2+n} (\delta q)_n (\frac{\gamma}{\alpha})_n}{(\gamma q)_n}$$

$$\sum_{k=0}^{n} \frac{a_2^k (-\alpha)^{-k} q^{k(k+1)/2+(k+1)(n-k)} (\gamma\delta q^{n+1})_k (q^{-n})_k (\alpha q)_k}{(q)_k (\alpha\delta q^{k+1})_n (1-\alpha\delta q^{n+k+1})}$$

$$(1-\alpha\delta q^{2k+1}) \frac{(\frac{\alpha\delta q^2}{b_1})_k (b_1/a_2)_k}{(\delta q)_k (1 - \frac{\gamma}{\alpha} q^{n-1})(1 - \frac{\gamma}{\alpha} q^{n-2}) \ldots (1- \frac{\gamma}{\alpha} q^{n-k}) (b_1)_k (\frac{\alpha\delta a_2 q^2}{b_1})_k}$$

$$= \frac{\alpha^n q^n (\delta q)_n (\gamma/\alpha)_n}{(\gamma q)_n (\alpha\delta q)_n} \sum_{k=0}^{n} \frac{a_2^k \gamma^{-k} (\gamma\delta q^{n+1})_k (q^{-n})_k (\alpha q)_k}{(q)_k (\alpha\delta q^{n+1})_k (1-\alpha\delta q^{n+k+1})}$$

$$\cdot (\alpha\delta q)_k (1-\alpha\delta q^{2k+1}) \cdot \frac{(\frac{\alpha\delta q^2}{b_1})_k (b_1/a_2)_k}{(\delta q)_k (\frac{\alpha}{\gamma} q^{-n+1})_k (b_1)_k (\frac{\alpha\delta a_2 q^2}{b_1})_k}$$

$$= \frac{\alpha^n q^n (\delta q)_n (\gamma/\alpha)_n}{(\alpha\delta q^2)_n (\gamma q)_n} \sum_{k=0}^{n} \frac{(\alpha\delta q)_k (1-\alpha\delta q^{2k+1}) (\frac{\alpha\delta q^2}{b_1})_k (\alpha q)_k}{(q)_k (1-\alpha\delta q) (b_1)_k (\delta q)_k}$$

$$\cdot \frac{(\gamma\delta q^{n+1})_k (b_1/a_2)_k (q^{-n})_k a_2^k \gamma^{-k}}{(\frac{\alpha}{\gamma} q^{-n+1})_k (\frac{\alpha\delta a_2 q^2}{b_1})_k (\alpha\delta q^{n+2})_k} . \tag{3.19}$$

In (3.19) we not set $\alpha = \frac{d}{q}$, $\gamma = eda^{-1}q^{-n-1}$, $\delta = \frac{a}{d}$, $b_1 = \frac{aq}{b}$, $a_2 = \frac{aq}{bc}$. Hence

$${}_4\phi_3 \left(\begin{matrix} q^{-n}, e, d, \frac{aq}{bc}; \ q, q \\ \frac{ed}{aq^n}, \frac{aq}{b}, \frac{aq}{c} \end{matrix} \right)$$

$$= \frac{d^n(\frac{aq}{d})_n(\frac{e}{aq^n})_n}{(aq)_n(\frac{ed}{aq^n})_n}$$

$$\cdot\ {}_8\phi_7\left(\begin{matrix}q,q\sqrt{a},-q\sqrt{a},b,d,e,c,q^{-n} & ;q,\frac{a^2q^{n+2}}{bcde}\\ \sqrt{a},-\sqrt{a},\frac{aq}{b},\frac{aq}{d},\frac{aq}{e},\frac{aq}{c},aq^{n+1} & \end{matrix}\right)$$

$$= \frac{(\frac{aq}{d})_n(\frac{aq}{e})_n}{(aq)_n(\frac{aq}{de})_n}\ {}_8\phi_7\left(\begin{matrix}a,q\sqrt{a},-q\sqrt{a},b,c,d,e,q^{-n} & ;q,\frac{a^2q^{n+2}}{bcde}\\ \sqrt{a},-\sqrt{a},\frac{aq}{b},\frac{aq}{c},\frac{aq}{d},\frac{aq}{e},aq^{n+1} & \end{matrix}\right), \quad (3.20)$$

which is equivalent to (3.18), the desired result.

The final result of this section Theorem 13 contains all the relations necessary for the proofs of (1.1), (1.2), and (1.1.1)-(1.1.6). All of these results were given by Rogers [20], [21]; although G.N. Watson [25], [26] was the first to see these as deducible from Theorem 12.

Theorem 13.

$$\frac{1}{(aq)_\infty}\left(1 + \sum_{n=1}^{\infty}\frac{(aq)_{n-1}(1-aq^{2n})(-1)^n a^{2n} q^{\frac{1}{2}n(5n-1)}}{(q)_n}\right)$$

$$= \sum_{n=0}^{\infty}\frac{q^{n^2}a^n}{(q)_n}$$

$$= (-aq^2;q^2)_\infty \sum_{n=0}^{\infty}\frac{q^{n^2}a^n}{(q^2;q^2)_n(-aq^2;q^2)_n}; \quad (3.21)$$

$$\sum_{n=0}^{\infty}\frac{q^{4n^2}q^{2n}}{(q^4;q^4)_n} = (aq;q^2)_\infty \sum_{n=0}^{\infty}\frac{q^{n^2}a^n}{(q^2;q^2)_n(aq;q^2)_n}. \quad (3.22)$$

Remark. The proof below of the first part of (3.21) is due to G.N. Watson [25]. The remaining two identities have been given several simple proofs (Rogers [20], Watson [26], Andrews [2]). The proofs we present rely on formulae for polynomial ${}_2\phi_1$-functions; while this appears more complicated it does provide us a good idea of where these identities come from, and it shows that all the results are related to the little q-Jacobi polynomials, namely the general polynomial ${}_2\phi_1$-function.

Proof. The first part of (3.21) follows from Theorem 12 by letting b, c, d, e, n all tend to infinity; rigorous justification of this procedure is given in the books by Bailey [11] and Slater [23].

To prove the second portion of (3.21) we call on the q-analog of Kummer's theorem:

$$ {}_2\phi_1\left(\begin{matrix} a,\ b;\ q,-q/b \\ \frac{aq}{b} \end{matrix}\right) = \frac{(aq;q^2)_\infty(-q)_\infty(q^2a/b^2;q^2)_\infty}{(aq/b)_\infty(-q/b)_\infty} . \tag{3.23} $$

This result - due to Bailey [12] and Daum [14] independently - can be proved in a very elementary manner [4]. In (3.23) we set $a = q^{-n}$ and then let $b \to 0$; this yields

$$ \sum_{j=0}^{n} \frac{(q^{-n})_j}{(q)_j} q^{nj-j(j-1)/2} = \begin{cases} 0 & \text{if } n \text{ is odd} \\ (q;q^2)_{n/2} & \text{if } n \text{ is even,} \end{cases} $$

a famous result of Gauss used in one of his treatments of the sign of the Gaussian sum. Continuing, we see that

$$ (aq;q^2) \sum_{n=0}^{\infty} \frac{q^{n^2}a^n}{(q^2;q^2)_n(aq;q^2)_n} $$

$$ = \sum_{n=0}^{\infty} \frac{q^{n^2}a^n}{(q^2;q^2)_n} (aq^{2n+1};q^2)_\infty $$

$$ = \sum_{n=0}^{\infty} \frac{q^{n^2}a^n}{(q^2;q^2)_n} \sum_{m=0}^{\infty} \frac{a^m q^{m^2+2nm}(-1)^m}{(q^2;q^2)_m} $$

$$ = \sum_{s=0}^{\infty} \frac{q^{s^2}a^s}{(q^2;q^2)_s} \sum_{m=0}^{s} \frac{(q^2;q^2)_s\ (-1)^m}{(q^2;q^2)_m(q^2;q^2)_{s-m}} $$

$$= \sum_{s=0}^{\infty} \frac{q^{s^2} a^s}{(q^2;q^2)_s} \sum_{m=0}^{s} \frac{(q^{-2s};q^2)_m}{(q^2;q^2)_m} q^{2sm-m^2+m}$$

$$= \sum_{s=0}^{\infty} \frac{q^{(2s)^2} q^{2s}}{(q^2;q^2)_{2s}} (q^2;q^4)_s \qquad \text{(by (3.24))}$$

$$= \sum_{s=0}^{\infty} \frac{q^{4s^2} q^{2s}}{(q^4;q^4)_s},$$

which establishes (3.22). We note that our proof follows Watson's approach [26]; however the fact that (3.23) implies (3.24) appears to be new.

Next we require a summation formula for the ${}_2\phi_1$-function which has probably not appeared before

$$
{}_2\phi_1\left(\begin{matrix} b^2, \frac{b^2}{c}; q^2, \frac{cq}{b^2} \\ c \end{matrix}\right)
= \frac{1}{2} \frac{(b^2;q^2)_\infty (q;q^2)_\infty}{(c;q^2)_\infty (cq/\beta^2;q^2)_\infty} \left(\frac{(c/b)_\infty}{(b)_\infty} + \frac{(-c/b)_\infty}{(-b)_\infty} \right).
$$

This result is easily deduced from Heine's fundamental transformation for the ${}_2\phi_1$ [2; (I1)]:

$$
{}_2\phi_1\left(\begin{matrix} \alpha, \beta \ ; q, \tau \\ \gamma \end{matrix}\right) = \frac{(\beta)_\infty (\alpha\tau)_\infty}{(\gamma)_\infty (\tau)_\infty} \, {}_2\phi_1\left(\begin{matrix} \gamma/\beta, \tau \ ; q, \beta \\ \alpha\tau \end{matrix}\right). \tag{3.26}
$$

To deduce (3.25), replace q by q^2 and set $\alpha = b^2/c$, $\beta = b^2$, $\gamma = c$, $\tau = cq/b^2$:

$$
{}_2\phi_1\left(\begin{matrix} b^2, b^2/c; q^2, cq/b^2 \\ c \end{matrix}\right)
$$

$$= \frac{(b^2;q^2)_\infty (q;q^2)_\infty}{(c;q^2)_\infty (cq/b^2;q^2)_\infty} {}_2\phi_1\left(\begin{matrix} c/b^2, cq/b^2; q^2, b^2 \\ q \end{matrix}\right)$$

$$= \frac{(b^2;q^2)_\infty (q;q^2)_\infty}{(c;q^2)_\infty (cq/b^2;q^2)_\infty} \sum_{n=0}^{\infty} \frac{(c/b^2)_{2n} b^{2n}}{(q)_{2n}}$$

$$= \frac{(b^2;q^2)_\infty (q;q^2)_\infty}{(c;q^2)_\infty (cq/b^2;q^2)_\infty} \frac{1}{2} \sum_{n=0} \frac{(c/b^2)_n b^n (1+(-1)^n)}{(q)_n}$$

$$= \frac{1}{2} \frac{(b^2;q^2)_\infty (q;q^2)_\infty}{(c;q^2)_\infty (cq/b^2;q^2)_\infty} \left(\frac{(c/b)_\infty}{(b)_\infty} + \frac{(-c/b)_\infty}{(-b)_\infty} \right) \quad \text{(by (3.6))},$$

which is (3.25). Now we set $b = q^{-n}$ in (3.25) and let $c \to \infty$; this yields

$$\sum_{m=0}^{n} \frac{(q^{-2n};q^2)}{(q^2;q^2)_m} q^{2nm-m^2+2m} = (-q)_n. \tag{3.26}$$

Thus to conclude the proof of (3.21) we see that

$$(-aq^2;q^2) \sum_{n=0}^{\infty} \frac{q^{n^2} a^n}{(q^2;q^2)_n (-aq^2;q^2)_n}$$

$$= \sum_{n=0}^{\infty} \frac{q^{n^2} a^n}{(q^2;q^2)_n} (-aq^{2n+2};q^2)_\infty$$

$$= \sum_{n=0}^{\infty} \frac{q^{n^2} a^n}{(q^2;q^2)_n} \sum_{m=0}^{\infty} \frac{q^{m^2+m+2nm} a^m}{(q^2;q^2)_m}$$

$$= \sum_{s=0}^{\infty} \frac{q^{s^2} a^s}{(q^2;q^2)_s} \sum_{m=0}^{s} \frac{(q^2;q^2)_s \; q^m}{(q^2;q^2)_m (q^2;q^2)_{s-m}}$$

$$= \sum_{s=0}^{\infty} \frac{q^{s^2} a^s}{(q^2;q^2)_s} \sum_{m=0}^{s} \frac{(q^{-2s};q^2)_m q^{2sm-m^2+2m}}{(q^2;q^2)_m}$$

$$= \sum_{s=0}^{\infty} \frac{q^{s^2} a^s}{(q^2;q^2)_s} (-q)_s$$

$$= \sum_{s=0}^{\infty} \frac{q^{s^2} a^s}{(q)_s} ,$$

as desired.

We now remark that our six identities (1.1.1)-(1.1.6) plus the Rogers-Ramanujan identities follow from Theorem 13. If we set $a = 1$ in Theorem 13 and recall that by Jacobi's triple product identity [18; p.284]:

$$\frac{\sum_{n=0}^{\infty} (-1)^n q^{\frac{1}{2}n(5n-1)} (1+q^n)}{(q)_\infty} = \prod_{n=0}^{\infty} \frac{1}{(1-q^{5n+1})(1-q^{5n+4})},$$

then (1.1), (1.1.5) and (1.1.1) follow respectively. If we set $a = q$ in (3.21) and now observe that [18; p.284]

$$\frac{\sum_{n=0}^{\infty} (-1)^n q^{\frac{1}{2}n(5n+3)} (1-q^{2n+1})}{(q)_\infty} = \prod_{n=0}^{\infty} \frac{1}{(1-q^{5n+2})(1-q^{5n+3})} ,$$

then (1.2) follows immediately and (1.1.4) follows after algebraic simplification.

Finally we note that the expression

$$(-aq;q^2) \sum_{n=0}^{\infty} \frac{q^{n^2+n} a^n}{(q^2;q^2)_n (-aq;q^2)_n}$$

is equal to each of the expressions in (3.21) since

$$(-aq;q^2)_\infty \sum_{n=0}^{\infty} \frac{q^{n^2+n}a^n}{(q^2;q^2)_n(-aq;q^2)_n}$$

$$= \sum_{n=0}^{\infty} \frac{q^{n^2+n}a^n}{(q^2;q^2)_n} (-aq^{2n+1};q^2)_\infty$$

$$= \sum_{n=0}^{\infty} \frac{q^{n^2+n}a^n}{(q^2;q^2)_n} \sum_{m=0}^{\infty} \frac{a^m q^{m^2+2nm}}{(q^2;q^2)_m}$$

$$= \sum_{m=0}^{\infty} \frac{q^{m^2}a^m}{(q^2;q^2)_m} \sum_{n=0}^{\infty} \frac{q^{n^2+n+2nm}a^n}{(q^2;q^2)_n}$$

$$= \sum_{m=0}^{\infty} \frac{q^{m^2}a^m}{(q^2;q^2)_m} (-aq^{2m+2};q^2)_\infty$$

$$= (-aq^2;q^2)_\infty \sum_{m=0}^{\infty} \frac{q^{m^2}a^m}{(q^2;q^2)_m(-aq^2;q^2)_m} .$$

Since the expression in (3.27) is now identified with those in (3.21) we see that (1.1.3) (in mild disguise) follows for $a = 1$ and (1.1.6) follows directly with $a = q$.

4. CONCLUSION

We have tried in the preceding pages to describe by example the interaction between the theory of partitions and the theory of basic hypergeometric functions. In particular, we hoped to emphasize the possible impact that q-orthogonal polynomials might have. We must add that the work presented here is the barest survey of our forthcoming treatise [9] on this topic, wherein q-analogs (sometimes more than one) of all the classical and discrete orthogonal polynomials are presented.

REFERENCES

1. G.E. Andrews, An analytic proof of the Rogers-Ramanujan-Gordon

identities, Amer. J. Math., 88(1966), 844-846.
2. G.E. Andrews, q-Identities of Auluk, Carlitz and Rogers, Duke Math. J., 33(1966), 575-582.
3. G.E. Andrews, Number Theory, W.B. Saunders, Philadelphia, 1971.
4. G.E. Andrews, On the q-analog of Kummer's theorem and applications, Duke Math. J., 40(1973), 525-528.
5. G.E. Andrews, A general theory of identities of the Rogers-Ramanujan type, Bull. Amer. Math. Soc., 80(1974), 1033-1052.
6. G.E. Andrews, An analytic generalization of the Rogers-Ramanujan identities for odd moduli, Proc. Nat. Acad. Sci., 71(1974), 4082-4085.
7. G.E. Andrews, Problems and propsects for basic hypergeometric functions, from Theory and Application of Special Functions, ed. R. Askey, Academic Press, New York, 1975, pp.191-224.
8. G.E. Andrews, The Theory of Partitions, Encyclopedia of Mathematics and Its Applications, Vol. 2, Addison-Wesley, Reading, 1977.
9. G.E. Andrews and R.A. Askey, The Classical and Discrete Orthogonal Polynomials and Their q-Analogs.
10. R.A. Askey, Orthogonal Polynomials and Special Functions, Regional Conf. Series in Appl. Math., Vol. 21, S.I.A.M., Philadephia, 1975.
11. W.N. Bailey, Generalized Hypergeometric Series, Cambridge University Press, Cambridge, 1935 (Reprinted: Hafner, New York, 1964).
12. W.N. Bailey, A note on certain q-identities, Quart. J. Math., 12(1941), 173-175.
13. W.G. Connor, Partition theorems related to some identities of Rogers and Watson, Trans. Amer. Math. Soc., 214(1975), 95-111.
14. J.A. Daum, The basic analog of Kummer's theorem, Bull. Amer. Math. Soc., 48(1942), 711-713.
15. B. Gordon, A combinatorial generalization of the Rogers-Ramanujan identities, Amer. J. Math., 83(1961), 393-399.
16. B. Gordon, Some continued fractions of the Rogers-Ramanujan type, Duke Math. J., 32(1965), 741-748.
17. G.H. Hardy, Ramanujan, Cambridge University Press, Cambridge, 1940 (Reprinted: Chelsea, New York).
18. G.H. Hardy and E.M. Wright, An Introduction to The Theory of Numbers, 4th ed., Oxford University Press, Oxford, 1960.
19. W. Hahn, Über Orthogonalpolynome, die q-Differenzengleichungen genügen, Math. Nach., 2(1949), 4-34.
20. L.J. Rogers, Second memoir on the expansion of certain infinite products, Proc. London Math. Soc., 25(1894), 318-343.
21. L.J. Rogers, On two theorems of combinatory analysis and some allied identities, Proc. London Math. Soc.(2), 16(1917), 315-336.
22. W.A. Al-Salam and A. Verma, Orthogonality preserving operators and q-Jacobi polynomials, to appear.
23. L.J. Slater, Generalized Hypergeometric Functions, Cambridge University Press, Cambridge, 1966.

24. G. Szego, Orthogonal Polynomials, Colloquium Publications, vol. 23, 3rd ed., American Mathematical Society, Providence, 1967.
25. G.N. Watson, A new proof of the Rogers-Ramanujan identities, J. London Math. Soc., 4(1929), 4-9.
26. G.N. Watson, A note on Lerch's functions, Quart. J. Math., Oxford Ser., 8(1937), 43-47.

DISTRIBUTIONS EULERIENNES ET MAHONIENNES SUR LE GROUPE DES PERMUTATIONS

Dominique Foata

Département de Mathématique, Université de Strasbourg,
7, rue René-Descartes, 67084 Strasbourg, France

1. INTRODUCTION

Les nombres eulériens $A_{n,k}$ $(n \geq 1,\ 1 \leq k \leq n)$ sont définis par la relation de récurrence

$$A_{1,1} = 1 \qquad A_{1,k} = 0 \qquad \text{pour } k \neq 1$$
$$A_{n,k} = k\,A_{n-1,k} + (n-k+1)\,A_{n-1,k-1} \quad \text{pour } n \geq 2 \text{ et } 1 \leq k \leq n \;. \tag{1}$$

La table 1 ci-dessous donne les premières valeurs de ces nombres.

n \ k	1	2	3	4	5	6
1	1					
2	1	1				
3	1	4	1			
4	1	11	11	1		
5	1	26	66	26	1	
6	1	57	302	302	57	1

Table 1 .

De façon équivalente, ces nombres sont définis par la formule, dite de Worpitzky

$$x^n = \sum_{1 \leq k \leq n} A_{n,k} \binom{x+k-1}{n} \qquad (n \geq 1)\ , \tag{2}$$

M. Aigner (ed.), Higher Combinatorics, 27-49.

dont l'inverse (au sens de Möbius) est donnée par

$$A_{n,k} = \sum_{0\le i\le k} (-1)^i (k-i)^n \binom{n+1}{i} \quad (0 \le k \le n) . \tag{3}$$

Pour tout entier $n \ge 1$ le polynôme

$$A_n(t) = \sum_{1\le k\le n} A_{n,k}\, t^{k-1}$$

est appelé <u>polynôme eulérien</u> (de degré $n-1$). La fonction génératrice exponentielle des polynômes $A_n(t)$ $(n \ge 1)$ a pour expression

$$1+\sum_{n\ge 1} (u^n/n!) A_n(t) = (1-t)/(-t+\exp(u(t-1))) ,$$

qu'on peut encore écrire sous la forme

$$1+\sum_{n\ge 1} (u^n/n!) A_n(t) = (1-\sum_{n\ge 1} (u^n/n!)(t-1)^{n-1})^{-1} . \tag{4}$$

On trouvera dans Carlitz (1959), Riordan [(1958), p. 38-39 & 214-216], ainsi que dans Foata-Schützenberger (1970), les propriétés arithmétiques et combinatoires de ces nombres, en particulier, l'équivalence des formules (1), (2), (3) et (4).

Les nombres <u>mahoniens</u> $B_{n,k}$ $(n \ge 1 ,\ 0 \le k \le n(n-1)/2)$ (on me pardonnera ce néologisme à la mémoire du grand combinatorialiste le major P. A. MacMahon) sont définis par l'identité

$$\sum_{0\le k\le n(n-1)/2} B_{n,k}\, q^k = \prod_{1\le i\le n} (1-q^i)/(1-q) \quad (n \ge 1) . \tag{5}$$

En introduisant le q-analogue $[i]_q$ de l'entier $i \ge 1$, à savoir

$$[i]_q = (1-q^i)/(1-q) = 1+q+q^2+\ldots+q^{i-1} , \tag{6}$$

on a ainsi

$$\sum_{0\le k\le n(n-1)/2} B_{n,k}\, q^k = [n]_q [n-1]_q \cdots [2]_q [1]_q \quad (n \ge 1).$$

Les premières valeurs de ces nombres sont données dans la table 2.

n \ k	0	1	2	3	4	5	6	7	8	9	10
1	1										
2	1	1									
3	1	2	2	1							
4	1	3	5	6	5	3	1				
5	1	4	9	15	20	22	20	15	9	4	1

<u>Table 2.</u>

On vérifie directement à partir des formules (1) et (5) que l'on a

$$\sum_{1\le k\le n} A_{n,k} = \sum_{0\le k\le n(n-1)/2} B_{n,k} = n! \quad (n \ge 1) .$$

On peut donc s'attendre à retrouver ces deux suites de nombres dans divers problèmes d'énumération concernant le groupe $\mathfrak{S}_n$ des permutations de l'intervalle $[n] = \{1, 2, \ldots, n\}$, ou tout autre ensemble fonctionnel de cardinal $n!$. Soient D_n un tel ensemble et X une application définie sur D_n à valeurs entières ; on dit que X est une statistique eulérienne (resp. mahonienne) sur D_n, si l'on a

$$A_{n,k} = \text{card}\,\{d \in D_n : X(d) = k\} \quad (1 \le k \le n)$$

$$(\text{resp. } B_{n,k} = \text{card}\{d \in D_n : X(d) = k\} \quad (0 \le k \le n(n-1)/2)) .$$

Soit $\sigma = \sigma(1)\,\sigma(2) \ldots \sigma(n)$ une permutation de la suite $1\,2 \ldots n$. On désigne par $\text{RISE}\,\sigma$ le nombre de montées de σ, c'est-à-dire, avec la convention $\sigma(0) = 0$, le nombre d'indices i tels que $0 \le i \le n-1$ et $\sigma(i) < \sigma(i+1)$. Le nombre d'inversions $\text{INV}\,\sigma$ de σ est défini comme le nombre de couples (i, j) tels que $1 \le i < j \le n$ et $\sigma(i) > \sigma(j)$. Enfin, l'indice majeur $\text{MAJ}\,\sigma$ est la somme de tous les entiers i tels que $1 \le i \le n-1$ et $\sigma(i) > \sigma(i+1)$.

On vérifiera, par exemple, que pour

$$\sigma = \begin{pmatrix} 1 & 2 & 3 & 4 & 5 & 6 & 7 & 8 & 9 \\ 6 & 4 & 9 & 7 & 2 & 5 & 8 & 1 & 3 \end{pmatrix},$$

on a

$$\text{RISE}\,\sigma = 5 , \quad \text{INV}\,\sigma = 23 \quad \text{et} \quad \text{MAJ}\,\sigma = 15 .$$

Il est bien connu (cf. par exemple Riordan (1958), p. 213-216) que RISE est une statistique eulérienne sur $\mathfrak{S}_n$. De même, on sait que INV et MAJ sont tous deux des statistiques mahoniennes sur $\mathfrak{S}_n$ (cf. par exemple, Comtet (1970), p. 81 et p. 108). Notant σ^{-1} la permutation inverse de σ dans le groupe $\mathfrak{S}_n$, on pose

$$\text{IRISE}\,\sigma = \text{RISE}\,\sigma^{-1} \quad \text{et} \quad \text{IMAJ}\,\sigma = \text{MAJ}\,\sigma^{-1} .$$

On obtient évidemment une nouvelle statistique eulérienne IRISE et une statistique mahonienne IMAJ.

Le présent travail comprend trois parties. Dans la première (la section 2 suivante), on signale un problème resté ouvert entre les interprétations "discrètes" des nombres eulériens et l'interprétation "continue" en termes de découpage du cube unité à n dimensions. On montre, en effet, analytiquement que le rapport $A_{n,k}/n!$

est encore le volume de la portion du cube unité à n dimensions comprise entre les plans $\Sigma x_i = k-1$ et $\Sigma x_i = k$. Il s'agit donc de trouver une preuve combinatoire de ce résultat. Le problème, exposé dans la section 2, avait été proposé aux participants du colloque de Berlin. Une solution simple et élégante trouvée par Richard Stanley est reproduite à la fin de cet article.

La seconde partie de l'article (sections 3 et 4) se veut une contribution à l'étude de la distribution du 5-vecteur (RISE, IRISE, INV, MAJ, IMAJ), c'est-à-dire à l'étude de la fonction génératrice multivariée

$$F(x_1, x_2, x_3, x_4, x_5) = \Sigma\, x_1^{\mathrm{RISE}\sigma}\, x_2^{\mathrm{IRISE}\sigma}\, x_3^{\mathrm{INV}\sigma}\, x_4^{\mathrm{MAJ}\sigma}\, x_5^{\mathrm{IMAJ}\sigma},$$

la somme étant étendue à l'ensemble de tous les σ dans $\mathfrak{S}_n$. Nous n'avons pas d'expression analytique pour cette fonction génératrice. En revanche, nous montrons que l'on connaît toutes les distributions marginales bivariées, c'est-à-dire la fonction génératrice bivariée de tout couple de statistiquesextrait de la suite (RISE, IRISE, INV, MAJ, IMAJ). Dans la section 3, au moyen d'une transformation sur $\mathfrak{S}_n$, déjà introduite dans Foata (1968) et employée dans Foata-Schützenberger (1977a), on montre que ces 20 distributions bivariées se réduisent à 6 distributions. La section 4 se borne à redonner l'expression analytique de ces 6 distributions On verra, en particulier, que le q-analogue des nombres eulériens trouvé par Carlitz [(1954), (1975)] donne la fonction génératrice bivariée de (RISE, MAJ), mais que le q-analogue des polynômes eulériens imaginé par Stanley (1976a) fournit celle de (IRISE, MAJ).

Pour tout $\sigma = \sigma(1)\,\sigma(2) \ldots \sigma(n)$ dans $\mathfrak{S}_n$ et $i = 1, 2, \ldots, n$, notons x_i le nombre de $\sigma(j)$ à gauche de $\sigma(i)$ qui sont inférieurs à $\sigma(i)$. Puis, désignons par IMALσ le nombre d'entiers distincts de la suite $x_1\, x_2 \ldots x_n$. Par exemple, pour

$$\begin{array}{rl} \sigma & = 7\ 9\ 1\ 2\ 8\ 5\ 6\ 3\ 4 \quad , \text{ on a} \\ x_1\, x_2 \ldots x_n & = 0\ 1\ 0\ 1\ 3\ 2\ 3\ 2\ 3\ . \end{array}$$

D'où IMAL$\sigma = 4$. Comme vérifié par Dumont (1974), la statistique IMAL est eulérienne. On démontre, dans la dernière partie de l'article, que le couple (IMAL, MAJ) a même distribution sur $\mathfrak{S}_n$ que (IRISE, MAJ) (et encore (RISE, INV)). Pour obtenir ce résultat, nous faisons appel à un codage des permutations, le S-code, dont les propriétés sont systématiquement exploitées dans un article en cours de rédaction (Foata-Schützenberger (1977b)).

Pour décrire les différents algorithmes utilisés, nous nous sommes permis d'utiliser les notations jugées traditionnelles en informatique. Par exemple "$x \leftarrow y$" signifie qu'il faut substituer y à x et on entend par "$x_1\, x_2 \ldots x_n$" $\leftarrow$ "$y_1\, y_2 \ldots y_n$" les substitutions $x_1 \leftarrow y_1$, $x_2 \leftarrow y_2, \ldots , x_n \leftarrow y_n$.

2. LES STATISTIQUES EULÉRIENNES

Pour démontrer que RISE, dont la définition a été donnée dans l'introduction, est une statistique eulérienne sur $\mathfrak{S}_n$, on vérifie que le nombre de permutations ayant k montées satisfait bien la relation de récurrence (1) . La démonstration est immédiate.

Pour tout $n \geq 1$, notons D_n l'ensemble des suites $x_1 x_2 \ldots x_n$, de longueur n , telles que $0 \leq x_i \leq i-1$ pour tout $i = 1, 2, \ldots, n$. Naturellement $\text{card}\, D_n = n!$. Pour chaque $w = x_1 x_2 \ldots x_n$ dans D_n on désigne par IMA w le nombre d'entiers distincts de la suite $x_1 x_2 \ldots x_n$. C'est Dumont (1974) qui a introduit cette statistique. A l'aide de la relation de récurrence (1) on vérifie immédiatement qu'elle est eulérienne. Dumont (1974) a construit une bijection Φ de $\mathfrak{S}_n$ sur D_n satisfaisant l'identité

$$\text{RISE}\, \sigma = \text{IMA}\, \Phi(\sigma) .$$

Il semble d'ailleurs que pour toutes les statistiques eulériennes E que l'on connaisse et qui sont définies sur des ensembles discrets, on sache construire une bijection Φ satisfaisant identiquement $\text{RISE}\, \sigma = E\, \Phi(\sigma)$ (cf. Foata-Schützenberger (1970)) .

Une dernière interprétation géométrique des nombres eulériens se rapportant à un ensemble continu et non plus à un ensemble discret restait jusqu'ici isolé combinatoirement des autres. Considérons le cube unité à n dimensions et pour $1 \leq k \leq n$ notons $T_{n,k}$ la partie de ce cube comprise entre les plans d'équation $\sum_{1 \leq i \leq n} x_i = k-1$ et $\sum_{1 \leq i \leq n} x_i = k$. Pour évaluer le volume de $T_{n,k}$ on peut procéder comme suit. Soit $X_1, X_2, \ldots, X_n$ une suite de n variables aléatoires, mutuellement indépendantes et uniformément réparties sur l'intervalle $[0,1]$. Posons $S_n = X_1 + X_2 + \ldots + X_n$. Le volume $\text{Vol}\, T_{n,k}$ est égal à la probabilité pour que S_n soit compris entre k-1 et k . Or il est élémentaire de déterminer la fonction de répartition $P\{S_n \leq x\}$ de S_n . Le calcul est fait dans Feller (1966) . Il est déjà implicite dans Laplace (1820). Pour tout x réel posons $x_+ = \max(x, 0)$. On a

$$P\{S_n \leq x\} = (1/n!) \sum_{0 \leq i \leq n} (-1)^i (x-i)_+^n \binom{n}{i} .$$

Lorsque x est égal à un entier k avec $0 \leq k \leq n$, on obtient

$$P\{S_n \leq k\} = (1/n!) \sum_{0 \leq i \leq k} (-1)^i (k-i)^n \binom{n}{i} .$$

D'où pour $1 \leq k \leq n$

$$\begin{aligned}\mathrm{Vol}\ T_{n,k} = P\{k-1<S_n\le k\} &= (1/n!)[\sum_{0\le i\le k} (-1)^i(k-i)^n\binom{n}{i} \\ &\qquad - \sum_{0\le i\le k-1} (-1)^i(k-1-i)\binom{n}{i}] \\ &= (1/n!) \sum_{0\le i\le k} (-1)^i(k-i)^n\binom{n+1}{i} \ .\end{aligned}$$

On retrouve donc l'expression (3) de Worpitzky pour les nombres eulériens. D'où

$$\mathrm{Vol}\ T_{n,k} = A_{n,k}/n! \quad (1 \le k \le n) \ . \qquad (6)$$

Le présent calcul est élémentaire, mais est refait périodiquement (cf. von Randow et al. (1971)). Remarquons que la démonstration de la formule (6) n'a été possible que grâce à la formule de Worpitzky. Le problème qui reste donc ouvert est de prouver (6) à partir de la formule de récurrence (1) , ou mieux, en établissant une correspondance biunivoque avec une statistique eulérienne connue. Nous formulons ce problème comme suit. Soit $U_{n,k}$ l'ensemble des points $(x_1, x_2, \dots, x_n)$ du cube unité ayant k montées, c'est-à-dire, avec la convention $x_0 = 0$, tels que l'on ait $x_{i-1} < x_i$ pour exactement k indices i pris dans $[n]$. Toujours à l'aide de la précédente suite de variables aléatoires $(X_1, X_2, \dots, X_n)$, on voit que le volume de $U_{n,k}$ est encore égal à la probabilité pour que le vecteur $(X_1, X_2, \dots, X_n)$ appartienne à $U_{n,k}$. Or on a $(X_1, X_2, \dots, X_n) \in U_{n,k}$ si et seulement s'il existe $\sigma \in \mathfrak{S}_n$ tel que $\mathrm{RISE}\,\sigma = k$ et $X_{\sigma^{-1}(1)} < X_{\sigma^{-1}(2)} < \dots < X_{\sigma^{-1}(n)}$ D'où

$$\begin{aligned}\mathrm{Vol}\ U_{n,k} &= \Sigma\,\{P\{X_{\sigma^{-1}(1)} < X_{\sigma^{-1}(2)} < \dots < X_{\sigma^{-1}(n)}\} : \mathrm{RISE}\,\sigma = k\}. \\ &= (1/n!)\ \mathrm{card}\{\sigma \in \mathfrak{S}_n : \mathrm{RISE}\,\sigma = k\} \\ &= (1/n!)\ A_{n,k} \qquad (1 \le k \le n) \ .\end{aligned}$$

Ainsi

$$\mathrm{Vol}\ T_{n,k} = \mathrm{Vol}\ U_{n,k} \qquad (1 \le k \le n) \ . \qquad (7)$$

Pour avoir une théorie combinatoire complète de toutes les interprétations des nombres eulériens, il suffit de construire une transformation

$$\Phi : (x_1, x_2, \dots, x_n) \to (y_1, y_2, \dots, y_n)$$

du cube unité qui envoie bijectivement le simplexe $U_{n,k}$ sur $T_{n,k}$ pour tout $k = 1, 2, \dots, n$.

Ce problème a été posé oralement lors du colloque de Berlin

et Stanley (1976b) a trouvé la transformation simple suivante : pour tout $i = 1, 2, \ldots, n$, on pose

$$\begin{aligned} y_i &= 1 + x_{i-1} - x_i & \text{si } x_{i-1} < x_i \\ &= x_{i-1} - x_i & \text{si } x_{i-1} > x_i \,. \end{aligned}$$

L'inverse de cette transformation est donnée par

$$x_i = - y_1 - \ldots - y_i + 1 + [y_1 + \ldots + y_i]$$

$(1 \le i \le n)$. Naturellement, la transformation et son inverse ne sont pas définies sur les faces des simplexes, c'est-à-dire l'ensemble des points ayant au moins deux coordonnées égales, mais ces ensembles sont de mesure nulle.

3. LES DISTRIBUTIONS BIVARIÉES

Le vecteur (RISE, IRISE, INV, MAJ, IMAJ) comporte deux statistiques eulériennes, RISE et IRISE, ainsi que trois statistiques mahoniennes, INV, MAJ et IMAJ. Parmi les vingt couples qu'on peut former à partir de ces cinq statistiques, on trouve

(I) deux couples dont les composantes sont des statistiques eulériennes (RISE, IRISE) et (IRISE, RISE) ;

(II) six couples dont les composantes sont des statistiques mahoniennes (MAJ, INV), (IMAJ, INV), (IMAJ, MAJ), (MAJ, IMAJ), (INV, IMAJ) et (INV, MAJ) ;

six couples dont la première composante est eulérienne, et la seconde mahonienne, qu'on va partager en deux classes (III) et (IV) :

(III) (RISE, MAJ) et (IRISE, IMAJ) ;

(IV) (RISE, INV), (IRISE, INV), (IRISE, MAJ), (RISE, IMAJ) ;

(V) (resp. (VI)) les deux (resp. quatre) couples obtenus en permutant les composantes dans les deux (resp. quatre) couples du groupe (III) (resp. (IV)).

THÉORÈME 1. Les couples de statistiques appartenant à un même groupe (I), (II), (III), (IV), (V) ou (VI) ont même distribution.

Pour démontrer ce théorème, nous allons faire appel aux propriétés d'une transformation ψ de $\mathfrak{S}_n$ décrite dans Foata (1968) et Foata-Schützenberger (1977b). Donnons tout d'abord la construction de ψ.

Algorithme pour ψ.
Soit $\sigma = \sigma(1)\ \sigma(2) \ldots \sigma(n)$ une permutation. Pour obtenir $\psi(\sigma) = \tau = \tau(1)\ \tau(2) \ldots \tau(n)$, on applique à σ la procédure suivante :

(1) si $n = 1$, alors $\tau \leftarrow \sigma$ et l'algorithme est terminé ; sinon $k \leftarrow n-1$; $\sigma_k \leftarrow \sigma(1)\ \sigma(2) \ldots \sigma(n-1)$ et $\tau(n) \leftarrow \sigma(n)$;

(2) si $k = 1$, poser $\tau(1)$ égal à l'unique lettre de σ_1 et $\psi(\sigma) = \tau \leftarrow \tau(1)\ \tau(2) \ldots \tau(n)$; l'algorithme est terminé ; sinon comparer le premier terme $\sigma_k(1)$ de σ_k avec $\tau(k+1)$; si $\sigma_k(1)$ est plus grand (resp. plus petit) que $\tau(k+1)$, couper σ_k avant chaque lettre plus grande (resp. plus petite) que $\tau(k+1)$;

(3) à l'intérieur de chaque compartiment déterminé par les coupures, déplacer la première lettre de la première à la dernière place ;

(4) poser σ_{k-1} égal au mot ainsi transformé après avoir supprimé la dernière lettre ;

(5) poser $\tau(k)$ égal à cette dernière lettre ;

(6) remplacer k par $k-1$ et retourner en (2).

EXEMPLE. Si on applique ψ à

$$\sigma = 6\ 4\ 9\ 7\ 2\ 5\ 8\ 1\ 3$$

on obtient successivement

$$\begin{array}{llr}
\sigma_8 = & 6\ |4\ |9\ |\ 7\ \ 2\ |5\ |8\ \ 1\ .\ 3 & = \tau(9) \\
\sigma_7 = & 6\ |4\ \ 9\ |\ 2\ |7\ |5\ |1\ .\ 8 & = \tau(8) \\
\sigma_6 = & 6\ |9\ |4\ |\ 2\ |7\ |5\ .\ 1 & = \tau(7) \\
\sigma_5 = & 6\ |9\ \ 4\ \ 2\ |7\ .\ 5 & = \tau(6) \\
\sigma_4 = & 6\ |4\ |2\ \ 9\ .\ 7 & = \tau(5) \\
\sigma_3 = & 6\ |4\ |9\ .\ 2 & = \tau(4) \\
\sigma_2 = & 6\ |4\ .\ 9 & = \tau(3) \\
\sigma_1 = & 6\ .\ 4 & = \tau(2) \\
 & 6 & = \tau(1) \\
\tau = \psi(\sigma) = & 6\ \ 4\ \ 9\ \ 2\ \ 7\ \ 5\ \ 1\ \ 8\ \ 3 &
\end{array}$$

Dans Foata (1968) on a démontré que ψ était une bijection de $\mathfrak{S}_n$ sur lui-même et satisfaisait l'identité

$$\mathrm{MAJ}\,\psi(\sigma) = \mathrm{INV}\,\sigma\ . \tag{8}$$

Dans l'exemple précédent

$$\mathrm{INV}\,\sigma = 23 = \mathrm{MAJ}\,\tau \,.$$

Dans Foata-Schützenberger (1977a) une autre propriété de la transformation ψ a été établie. Notons RISE SETσ l'ensemble des montées d'une permutation $\sigma = \sigma(1)\,\sigma(2)\,\ldots\,\sigma(n)$, c'est-à-dire l'ensemble des entiers i tels que $0 \le i \le n-1$ et $\sigma(i) < \sigma(i+1)$ (avec toujours la convention $\sigma(0) = 0$). On pose aussi

$$\mathrm{IRISE\ SET}\,\sigma = \mathrm{RISE\ SET}\,\sigma^{-1} \,,$$

de sorte que RISEσ et IRISEσ sont respectivement les cardinaux de RISE SETσ et IRISE SETσ. Par ailleurs MAJ (resp. IMAJ) est la somme des entiers compris entre 1 et $n-1$ qui n'appartiennent pas à RISE SETσ (resp. IRISE SETσ). Il est immédiat que i appartient à IRISE SETσ si et seulement si $i+1$ est à la droite de i dans le mot $\sigma(1)\,\sigma(2)\,\ldots\,\sigma(n)$.

Par exemple, avec

$$\sigma = \begin{matrix} 0 \\ 0 \end{matrix}\begin{pmatrix} 1 & 2 & 3 & 4 & 5 & 6 & 7 & 8 & 9 \\ 6 & 4 & 9 & 7 & 2 & 5 & 8 & 1 & 3 \end{pmatrix}$$

on a RISE SET$\sigma = \{0, 2, 5, 6, 8\}$
et IRISE SET$\sigma = \{0, 2, 4, 6, 7\}$.

On démontre (cf. Foata-Schützenberger (1977a)) que l'on a identiquement dans $\mathfrak{S}_n$

$$\mathrm{IRISE\ SET}\,\psi(\sigma) = \mathrm{IRISE\ SET}\,\sigma \,.$$

On en déduit

$$\begin{aligned} \mathrm{IRISE}\ \psi(\sigma) &= \mathrm{IRISE}\,\sigma \\ \mathrm{IMAJ}\ \psi(\sigma) &= \mathrm{IMAJ}\,\sigma \end{aligned} \tag{9}$$

Posons $\underset{\sim}{i}\,\sigma = \sigma^{-1}$ pour tout σ dans $\mathfrak{S}_n$ et considérons la suite

$$\sigma \xrightarrow{\underset{\sim}{i}} \sigma_1 \xrightarrow{\psi} \sigma_2 \xrightarrow{\underset{\sim}{i}} \sigma_3 \xrightarrow{\psi^{-1}} \sigma_4 \xrightarrow{\underset{\sim}{i}} \sigma_5 \tag{10}$$

On déduit de (8) et de (9), et du fait bien connu que $\underset{\sim}{i}$ conserve le nombre des inversions d'une permutation, les égalités

$$\mathrm{MAJ}\sigma = \mathrm{IMAJ}\sigma_1 = \mathrm{IMAJ}\sigma_2 = \mathrm{MAJ}\sigma_3 = \mathrm{INV}\sigma_4 = \mathrm{INV}\sigma_5$$
$$\mathrm{INV}\,\sigma = \mathrm{INV}\ \sigma_1 = \mathrm{MAJ}\,\sigma_2 = \mathrm{IMAJ}\sigma_3 = \mathrm{IMAJ}\sigma_4 = \mathrm{MAJ}\sigma_5$$
$$\mathrm{RISE}\sigma = \mathrm{IRISE}\sigma_1 = \mathrm{IRISE}\sigma_2 \,; \qquad \mathrm{RISE}\sigma_4 = \mathrm{IRISE}\sigma_5$$

La suite (10) donne toutes les transformations utiles pour

établir le théorème 1 .

Q. E. D.

Les distributions des couples des groupes (V) et (VI) se ramènent, par symétrie, à celles des couples des groupes (III) et (IV) (respectivement). Il n'y a donc que quatre distributions marginales bivariées distinctes, celles des groupes (I), (II), (III) et (IV) . Leur expression analytique est connue, comme nous allons le voir dans la prochaine section.

4. EXPRESSIONS ANALYTIQUES DES DISTRIBUTIONS MARGINALES

Carlitz et al. (1966) ont obtenu plusieurs expressions analytiques pour la distribution de (RISE, IRISE) (groupe (I)) . Aux quatre formules (1), (2), (3) et (4) existant pour les nombres eulériens, ils ont pu associer quatre formules pour les nombres

$$A_{n,j,k} = \text{card}\{\sigma \in \mathfrak{S}_n : \text{RISE}\sigma = j,\ \text{IRISE}\sigma = k\} .$$

D'après Carlitz et al. (1966), aux formules (1), (2), (3) et (4) correspondent les formules (11), (12), (13) et (14) suivantes :

$$(n+1)A_{n+1,j,k} = (jk+n)A_{n,j,k} + (k(n+2-j)-n)A_{n,j-1,k} + (j(n+2-k)-n)A_{n,j,k-1} + ((n+2-j)(n+2-k)+n)\,A_{n,j-1,k-1} \tag{11}$$

$$\binom{xy+n-1}{n} = \sum_{0\le j,k\le n} \binom{x+j-1}{n}\binom{y+k-1}{n} A_{n,j,k} \tag{12}$$

$$A_{n,j,k} = \sum_{0\le s\le j}\ \sum_{0\le t\le k} (-1)^{s+t}\binom{n+1}{s}\binom{n+1}{t}\binom{(j-s)(k-t)+n-1}{n} \tag{13}$$

$$\sum_{n\ge 0}\ \sum_{j,k\ge 0} A_{n,j,k}\ x^j y^k u^n (1-x)^{-n}(1-y)^{-n} = \tag{14}$$

$$= \sum_{j\ge 0} y^j/(1-x(1-u)^{-j}) = \sum_{k\ge 0} x^k/(1-y(1-u)^{-k}) .$$

Considérons l'identité

$$\prod_{m,n\ge 0} (1-x^n y^m u) = \sum_{n\ge 0} H_n(x,y)u^n/((1-x)(1-y)\ldots(1-x^n)(1-y^n))$$

D'après Carlitz (1956), cette identité définit une suite de polynômes

$H_n(x,y)$ $(n \geq 0)$. Cheema et Motzkin (1971), et de nouveau Roselle (1974) ont montré que pour tout $n \geq 1$ le polynôme $H_n(x,y)$ était le polynôme générateur de (MAJ, IMAJ) sur $\mathfrak{S}_n$. D'après le théor. 1 on conclut que $H_n(x,y)$ est aussi le polynôme générateur de chacun des couples du groupe (II), en particulier de (MAJ, INV).

Carlitz (1954, 1975) a considéré le q-analogue de la relation de récurrence (1) des nombres eulériens. En remplaçant donc l'entier k par son q-analogue $[k]_q = (1-q^k)/(1-q)$, on définit une suite de polynômes $A_{n,k}(q)$ définis par

$$A_{1,1}(q) = 1 \qquad A_{1,k}(q) = 0 \qquad \text{pour } k \neq 1$$

$$A_{n,k}(q) = [k]_q \, A_{n-1,k}(q) + [n-k+1]_q \, A_{n-1,k-1}(q) \qquad \text{pour } n \geq 2$$

et $1 \leq k \leq n$.
Posant

$$A_n(t,q) = \sum_{1 \leq k \leq n} t^k \, q^{\binom{n-k+1}{2}} A_{n,k}(q) \quad (n \geq 1) ,$$

Carlitz (1975) a montré ensuite que $A_n(t,q)$ était le polynôme générateur de (RISE, MAJ) sur $\mathfrak{S}_n$. Une table des premières valeurs est donnée à la fin de cette section.

Par ailleurs, Stanley (1976a) a considéré le q-analogue non pas de la formule (1), mais de la formule (4) qui donne l'expression de la fonction génératrice des polynômes eulériens. Remplaçant dans cette formule $n!$ par son q-factoriel

$$[n]_q! = \frac{1-q^n}{1-q} \, \frac{1-q^{n-1}}{1-q} \cdots \frac{1-q}{1-q} ,$$

on obtient l'identité

$$1 + \sum_{n \geq 1} A'_n(t,q)\, u^n/[n]_q! = (1 - \sum_{n \geq 1} (t-1)^{n-1} u^n/[n]_q!)^{-1} ,$$

qui définit une suite de polynômes $(A'_n(t,q))_{n \geq 1}$. (On trouvera les premières valeurs ci-après.)
Stanley (1976) établit que $tA'_n(t,q)$ est le polynôme générateur de (RISE, INV) sur $\mathfrak{S}_n$. D'après le théorème 1, $tA'_n(t,q)$ est donc aussi le polynôme générateur de chacun des couples du groupe (IV), en particulier, de (IRISE, MAJ).

D'après le théorème 1 et les résultats de cette section, nous avons donc une expression pour la fonction génératrice bivariée de tous les couples de statistiques de la suite

(RISE, IRISE, INV, MAJ, IMAJ) .

$$A_1(t,q) = t$$
$$A_2(t,q) = tq+t^2$$
$$A_3(t,q) = tq^3+t^2q(2q+2)+t^3$$
$$A_4(t,q) = tq^6+t^2q^3(3q^2+5q+3)+t^3q(3q^2+5q+3)+t^4$$
$$A_5(t,q) = tq^{10}+t^2q^6(4q^3+9q^2+9q+4)+t^3q^3(6q^4+16q^3+22q^2+16q+6) +t^4q(4q^3+9q^2+9q+4)+t^5$$

Table des $A_n(t,q)$.

$$tA'_1(t,q) = t$$
$$tA'_2(t,q) = tq+t^2$$
$$tA'_3(t,q) = tq^3+t^2q(2q+2)+t^3$$
$$tA'_4(t,q) = tq^6+t^2(3q^5+4q^4+3q^3+q^2)+t^3(q^4+3q^3+4q^2+3q)+t^4$$
$$tA'_5(t,q) = tq^{10}+t^2(4q^9+6q^8+6q^7+6q^6+2q^5+2q^4) +t^3(3q^8+9q^7+12q^6+18q^5+12q^4+9q^3+3q^2) +t^4(2q^6+2q^5+6q^4+6q^3+6q^2+4q)+t^5$$

Table des $tA'_n(t,q)$.

Aucune fonction génératrice marginale trivariée ou quadrivariée de la distribution de ce vecteur semble connue. Notons que Foata et Schützenberger (1977a) ont étudié les symétries de la distribution du 4-vecteur (RISE, IRISE, MAJ, IMAJ) et montré que le groupe de symétrie sous-jacent était le produit direct du groupe dihédral d'ordre 8 par le groupe d'ordre 2 .

D'après le théorème 1, les couples (RISE, INV) et (IRISE, MAJ) ont même distribution. On ne connaît pas de statistique eulérienne E telle que (E, INV) et (RISE, MAJ) aient même distribution.

5. LE V-CODE.

La statistique eulérienne IMAL sur $\mathfrak{S}_n$ a été définie dans l'introduction. On se propose d'établir dans les trois dernières sections que les <u>couples (IMAL, MAJ) et (IRISE, MAJ) ont même distribution sur $\mathfrak{S}_n$</u> . Pour ce faire, nous définissons un V-code de $\mathfrak{S}_n$, puis un S-code (dans la section 6), qui est un réarrangement du V-code. Enfin, le résultat annoncé sera établi comme corollaire des propriétés du S-code, en section 7 .

Le <u>V-code</u> de $\sigma = \sigma(1)\,\sigma(2)\ldots\sigma(n)$ est un mot $V\sigma = v_1 v_2 \ldots v_n$ tel que pour chaque $i = 1, 2, \ldots, n$ la lettre v_i est égale à la position du plus grand entier $\sigma(j)$, inférieur à $\sigma(i)$ et situé à la gauche de $\sigma(i)$ (on convient toujours que $\sigma(0) = 0$ est en position 0). En d'autres termes, si $a_i = \max\{\sigma(j) : 0 \le j \le i-1, \sigma(j) < \sigma(i)\}$, alors v_i est l'unique entier tel que $\sigma(v_i) = a_i$.

La définition du V-code s'étend aux sous-mots des permutations $\sigma(1)\,\sigma(2)\ldots\sigma(n)$, c'est-à-dire aux suites $\sigma' = \sigma(c_1)\sigma(c_2)\ldots\sigma(c_k)$ telles que $k \ge 1$ et $1 \le c_1 < c_2 < \ldots < c_k \le n$. En posant $\sigma(c_0) = 0$ et $a'_i = \max\{\sigma(c_j) : 0 \le j \le i-1, \sigma(c_j) < \sigma(c_i)\}$, le V-code de σ' est $V\sigma' = v_1 v_2 \ldots v_k$ avec v_i l'unique entier tel que $\sigma(c_{v_i}) = a'_i$.

Par exemple avec

$$\begin{matrix} 0 & \\ \sigma = & \end{matrix}\begin{pmatrix} 1 & 2 & 3 & 4 & 5 & 6 & 7 & 8 & 9 \\ 5 & 7 & 1 & 4 & 6 & 3 & 9 & 2 & 8 \end{pmatrix}$$

on a

$$V\sigma = \quad 0 \; 1 \; 0 \; 3 \; 1 \; 3 \; 2 \; 3 \; 2 \; .$$

Pour le sous-mot

$$\begin{matrix} 0 & \\ \sigma' = & \end{matrix}\begin{pmatrix} 2 & 4 & 5 & 8 & 9 \\ 7 & 4 & 6 & 2 & 8 \end{pmatrix}$$

on a

$$V\sigma' = \quad 0 \; 0 \; 2 \; 0 \; 1 \; .$$

Rappelons qu'une avance de σ est un entier i tel que $0 \le i \le n-1$ et tel que $\sigma(i)+1$ est à la droite de $\sigma(i)$. On note $A\sigma$ l'ensemble des avances de σ. Naturellement $\mathrm{IRISE}\,\sigma = \mathrm{card}\, A\sigma$.

Par exemple, avec

$$\sigma = \begin{matrix} 0 \\ 0 \end{matrix}\begin{pmatrix} 1 & 2 & 3 & 4 & 5 & 6 & 7 & 8 & 9 \\ 5 & 7 & 1 & 4 & 6 & 3 & 9 & 2 & 8 \end{pmatrix}$$

on a $\quad A\sigma = \{0, 1, 2, 3\}$.

Il est clair que si σ est dans $\mathfrak{S}_n$, alors $V\sigma$ appartient à l'ensemble D_n de tous les mots $x_1 x_2 \ldots x_n$ tels que $0 \le x_i \le i-1$ pour tout $i = 1, 2, \ldots, n$. De plus, l'ensemble des lettres distinctes de $V\sigma$ est égal à l'ensemble des avances de σ. Comme le cardinal de $A\sigma$ est encore égal à $\mathrm{IRISE}\,\sigma$, on a

$$\mathrm{IMA}\, V\sigma = \mathrm{IRISE}\,\sigma \; . \tag{15}$$

Pour se convaincre que $V : \mathfrak{S}_n \to D_n$ est bien bijectif, on donne tout de suite la construction de l'inverse V^{-1} de V.

Algorithme pour V^{-1}. Soit $w = x_1 x_2 \ldots x_n$ un mot de D_n. Pour obtenir la permutation σ telle que $V\sigma = w$, on applique la procédure suivante :

(1) $\sigma(0) \leftarrow 0$; $\sigma(1) \leftarrow 1$; si $n = 1$, l'algorithme est terminé ; sinon $k \leftarrow 2$;

(2) $\sigma(k) \leftarrow \sigma(x_k)+1$ et incrémenter de 1 tous les éléments de la suite $\sigma(0)\,\sigma(1)\,\sigma(2)\ldots\sigma(k-1)$ qui sont supérieurs ou égaux à $\sigma(k)$ et laisser invariants les autres ;

(3) si $k = n$, l'algorithme est terminé ; sinon remplacer k par $k+1$ et aller en (2) .

Les deux propriétés suivantes du V-code sont essentielles pour la construction du S-code.

PROPRIÉTÉ 1. Soit $V\sigma = v_1 v_2 \ldots v_n$ le V-code d'une permutation $\sigma = \sigma(1)\,\sigma(2)\ldots\sigma(n)$. On pose $x = \max\{v_i : 1 \le i \le n\}$ et on note $(c_1, c_2, \ldots, c_k)$ (resp. $(d_1, d_2, \ldots, d_\ell)$) la suite croissante des entiers m tels que $v_m < x$ (resp. $v_m = x$) . Alors

$$\sigma(d_1) = \sigma(x)+\ell+1,\ \sigma(d_2) = \sigma(x)+\ell,\ \sigma(d_3) = \sigma(x)+\ell-1, \ldots,$$
$$\sigma(d_{\ell-1}) = \sigma(x)+2,\ \sigma(d_\ell) = \sigma(x)+1 .$$

De plus, le V-code du sous-mot $\sigma' = \sigma(c_1)\,\sigma(c_2)\ldots\sigma(c_k)$ est égal à $V\sigma' = v_{c_1} v_{c_2} \ldots v_{c_k}$.

La vérification de la propriété 1 ne présente aucune difficulté et n'est pas reproduite ici. Illustrons cette propriété avec l'exemple donné précédemment. On a

$$\begin{array}{rcccccccccc}
 & 0 & 1 & 2 & 3 & 4 & 5 & 6 & 7 & 8 & 9 \\
\sigma = & & 5 & 7 & 1 & 4 & 6 & 3 & 9 & 2 & 8 \\
V\sigma = & & 0 & 1 & 0 & 3 & 1 & 3 & 2 & 3 & 2 \\
x = & & 3 \\
c_1\ldots c_k = & & 1 & 2 & 3 & . & 5 & . & 7 & . & 9 \\
\sigma' = & & 5 & 7 & 1 & . & 6 & . & 9 & . & 8 \\
\sigma(x) = & & 1 \\
d_1\ldots d_\ell = & & . & . & . & 4 & . & 6 & . & 8 & . \\
\sigma(d_1)\ldots\sigma(d_\ell) = & & . & . & . & 4 & . & 3 & . & 2 & . \\
\text{et}\quad V\sigma' = & & 0 & 1 & 0 & . & 1 & . & 2 & . & 2 ,
\end{array}$$

qui est égal à $v_{c_1} v_{c_2} \ldots v_{c_k}$.

La propriété suivante est une conséquence immédiate de la propriété 1 .

PROPRIÉTÉ 2. Soient w et w' deux mots de D_n ne différant que par la position de leurs lettres maximales. En d'autres termes, supposons

$$w = v_1 \ldots v_{d_1-1}\, x\, v_{d_1+1} \ldots v_{d_\ell-1}\, x\, v_{d_\ell+1} \ldots v_n$$

$$w' = v'_1 \dots v'_{d'_1-1}\, x\, v'_{d'_1+1} \dots v'_{d'_\ell-1}\, x\, v'_{d'_\ell+1} \dots v'_n ,$$

avec

$$v_1 \dots v_{d_1-1} v_{d_1+1} \dots v_{d_\ell-1} v_{d_\ell+1} \dots v_n$$
$$= v'_1 \dots v'_{d'_1-1} v'_{d'_1+1} \dots v'_{d'_\ell-1} v'_{d'_\ell+1} \dots v'_n ,$$

ce dernier mot ne contenant que des lettres inférieures à x. Alors, avec $\sigma = V^{-1}w$ et $\sigma' = V^{-1}w'$, on a

$$\sigma(d_1) = \sigma'(d'_1),\ \sigma(d_2) = \sigma'(d'_2), \dots, \sigma(d_\ell) = \sigma'(d'_\ell) .$$

6. LE S-CODE.

Pour tout $w = x_1 x_2 \dots x_n$ appartenant à D_n, on note RISE SETw l'ensemble des montées de w, c'est-à-dire des indices i tels que $0 \le i \le n-1$ et $x_i < x_{i+1}$ (par convention : $x_0 = 0$). D'après (15) on a IMA $V\sigma$ = IRISEσ. En revanche RISE SET $V\sigma$ n'est pas forcément égal à RISE SETσ. Pour rétablir cette propriété, il suffit d'obtenir un réarrangement approprié de $V\sigma$, la propriété (15) étant évidemment conservée par tout réarrangement. Nous nous proposons donc d'établir le théorème suivant

THÉORÈME 2. *Il existe une bijection* $S : \mathfrak{S}_n \to D_n$ *telle que si*

$$S\sigma = s_1 s_2 \dots s_n \quad \text{et} \quad V\sigma = v_1 v_2 \dots v_n ,$$

on a les propriétés suivantes

(i) $S\sigma$ *est un réarrangement de* $V\sigma$
(ii) RISE SET $S\sigma$ = RISE SET σ.

Construction du S-code. On procède par récurrence sur n. Pour $n = 1$ on pose évidemment $S\sigma = V\sigma = 0$ pour l'unique permutation σ dans $\mathfrak{S}_1$. Soient $n \ge 2$ et $V\sigma = v_1 v_2 \dots v_n$ le V-code d'une permutation $\sigma = \sigma(1)\,\sigma(2) \dots \sigma(n)$. On pose $x = \max\{v_i : 1 \le i \le n\}$ et on note $(c_1, c_2, \dots, c_k)$ (resp. $(d_1, d_2, \dots, d_\ell)$) la suite croissante des indices m tels que $v_m < x$ (resp. $v_m = x$). Soit $t_{c_1} t_{c_2} \dots t_{c_k}$ le S-code de $\sigma(c_1)\,\sigma(c_2) \dots \sigma(c_k)$. On forme le mot

$$w = t_1 t_2 \dots t_n$$
$$= t_1 \dots t_{d_1-1}\, x\, t_{d_1+1} \dots t_{d_2-1}\, x\, t_{d_2+1} \dots t_{d_\ell-1}\, x\, t_{d_\ell+1} \dots t_n .$$

Le S-code $s_1 s_2 \dots s_n$ de $\sigma = \sigma(1)\,\sigma(2) \dots \sigma(n)$ est obtenu par la procédure suivante :

(1) $s_1 s_2 \dots s_{d_1-1} \leftarrow t_1 t_2 \dots t_{d_1-1}$;

$\sigma(n+1) \leftarrow 0$; $d_{\ell+1} \leftarrow n+1$; $j \leftarrow 1$;

(2) $m \leftarrow d_j$;

(3) si $\sigma(m) > \sigma(m-1)$ et $\sigma(m+1)$, alors

$$s_m\, s_{m+1} \dots s_{d_{j+1}-1} \leftarrow x\, t_{m+1} \dots t_{d_{j+1}-1}$$

et aller en (6) ;

(4) si $\sigma(m) < \sigma(m-1) < \sigma(m+1)$ et $s_{m-1} < t_{m+1}$, ou si $\sigma(m-1) > \sigma(m) > \sigma(m+1)$, noter q la position du pic le plus voisin de m et situé à sa gauche ; autrement dit, q est défini par les inégalités

$$1 \le q \le m-1,\ \sigma(q-1) < \sigma(q),\ \sigma(q) > \sigma(q+1) > \dots > \sigma(m-1) > \sigma(m) ;$$

alors

$$s_q s_{q+1} s_{q+2} \dots s_{m-1} s_m s_{m+1} \dots s_{d_{j+1}-1}$$
$$\leftarrow x\, s_q s_{q+1} \dots s_{m-2} s_{m-1} t_{m+1} \dots t_{d_{j+1}-1}$$

et aller en (6) ;

(5) (dans les autres cas) noter q la position du pic le plus voisin de m et situé à sa droite ; autrement dit, q est l'entier défini par $m+1 \le q \le n$,

$$\sigma(m) < \sigma(m+1) < \dots < \sigma(q) \quad \text{et} \quad \sigma(q) > \sigma(q+1) ;$$

alors

$$s_m s_{m+1} \dots s_{q-1} s_q s_{q+1} \dots s_{d_{j+1}-1}$$
$$\leftarrow t_{m+1} t_{m+2} \dots t_q x t_{q+1} \dots t_{d_{j+1}-1} ;$$

(6) si $j = \ell$ l'algorithme est terminé et on pose

$$S\sigma \leftarrow s_1 s_2 \dots s_n ;$$

sinon incrémenter j de 1 et aller en (2) .

REMARQUE 1. L'instruction (5) de l'algorithme précédent est bien définie. En effet, on a $\sigma(m) = \sigma(d_j) > \sigma(d_{j+1})$ d'après la propriété 1 . Les inégalités $\sigma(m) < \sigma(m+1) < \dots < \sigma(q)$ entraînent donc $(m=)d_j < q < d_{j+1}$.

REMARQUE 2. On vérifie que $S\sigma$ appartient bien à D_n en s'assurant qu'à chaque étape, les s_i que l'on définit satisfont $s_i \le i-1$.

REMARQUE 3. Le fait que

$$\text{RISE SET } S\sigma = \text{RISE SET } \sigma$$

se vérifie sans difficulté lors de l'application de chaque instruction (3), (4) et (5).

REMARQUE 4. Si le V-code d'une permutation σ ne contient que des lettres au plus égales à 1, on a $S\sigma = V\sigma$.

REMARQUE 5. Soient $V\sigma = v_1 v_2 \dots v_n$ le V-code d'une permutation $\sigma = \sigma(1)\,\sigma(2) \dots \sigma(n)$ et $x = \max\{v_i : 1 \le i \le n\}$. Pour chaque $i = 1, 2, \dots, x$, on note $c_1^i, c_2^i, \dots, c_{ki}^i$ la suite croissante des indices m tels que $v_m \le i$ et σ_i le sous-mot $\sigma_i = \sigma(c_1^i)\,\sigma(c_2^i) \dots \sigma(c_{ki}^i)$. Pour appliquer l'algorithme précédent, on détermine successivement $S\sigma_i$ pour chaque $i = 1, 2, \dots, x$.

Dans l'exemple suivant, on a indiqué la suite des instructions de l'algorithme (notées de I1 à I6) seulement pour la détermination de $S\sigma_3$.

	1	2	3	4	5	6	7	8	9	10	11	12	13	14	
$\sigma =$	9	11	3	10	2	8	7	14	13	6	5	1	4	12	
$V\sigma =$	0	1	0	1	0	3	3	2	2	3	3	0	3	2	
$S\sigma_1 =$	0	1	0	1	0	.	.	.	.	.	.	0	.	.	
$S\sigma_2 =$	0	1	0	1	0	.	.	2	2	.	.	0	.	2	
$w =$	0	1	0	1	0	3	3	2	2	3	3	0	3	2	
$s_1 \dots s_5 =$	0	1	0	1	0										(I1)
$s_1 \dots s_6 =$	0	1	0	1	0	3									(I3)
$s_1 \dots s_9 =$	0	1	0	1	0	3	2	3	2						(I5)
$s_1 \dots s_{10} =$	0	1	0	1	0	3	2	3	3	2					(I4)
$s_1 \dots s_{11} =$	0	1	0	1	0	3	2	3	3	3	2	0			(I4)
$S\sigma =$	0	1	0	1	0	3	2	3	3	3	2	0	2	3	(I5)

Nous donnons ci-après la construction de l'inverse S^{-1} de S. Il est fastidieux mais facile de vérifier que $S^{-1} \circ S$ est l'application identique. Dans la description de S^{-1} on verra que la propriété 2 du V-code joue un rôle crucial.

Algorithme pour S^{-1}. Soient $w = s_1 s_2 \dots s_n$ un mot de D_n et $x = \max\{s_i : 1 \le i \le n\}$. On note $d_1, d_2, \dots, d_\ell$ les indices i tels que $s_i = x$. Le mot $w' = s_1 \dots s_{d_1-1} s_{d_1+1} \dots s_{d_\ell-1} s_{d_\ell+1} \dots s_n$ est, par récurrence, le S-code d'une permutation σ' dont le V-code est un réarrangement de w', qu'on note

$$v' = v_1 \dots v_{d_1 - 1} v_{d_1 + 1} \dots v_{d_\ell - 1} v_{d_\ell + 1} \dots v_n \ .$$

Formons le mot

$$\begin{aligned} v &= v_1 v_2 \dots v_n \\ &= v_1 \dots v_{d_1 - 1} x\, v_{d_1 + 1} \dots v_{d_\ell - 1} x\, v_{d_\ell + 1} \dots v_n \ , \end{aligned}$$

et déterminons la permutation

$$\tau = \tau(1)\ \tau(2) \dots \tau(n) \text{ dont le V-code est } v \ .$$

Pour obtenir la permutation $\sigma = \sigma(1)\ \sigma(2) \dots \sigma(n)$ dont le S-code est w, on applique la procédure suivante :

(1) $\sigma(d_\ell+1)\ \sigma(d_\ell+2) \dots \sigma(n) \leftarrow \tau(d_\ell+1)\ \tau(d_\ell+2) \dots \tau(n)$;
$\sigma(0) \leftarrow 0$; $\sigma(n+1) \leftarrow 0$; $d_0 \leftarrow 0$; $j \leftarrow \ell$;
(2) $m \leftarrow d_j$;
(3) si $\tau(m) > \tau(m-1)$ et $\sigma(m+1)$, alors

$$\sigma(d_{j-1}+1)\sigma(d_{j-1}+2)\dots\sigma(m-1)\sigma(m) \leftarrow \tau(d_{j-1}+1)\tau(d_{j-1}+2)\dots\tau(m-1)\tau(m)$$

et aller en (6) ;
(4) si $\tau(m-1)$ et $\tau(m) < \sigma(m+1)$,
on note p le plus petit indice tel que $m+1 \le p \le n$ et
$\sigma(m+1) > \sigma(m+2) > \dots > \sigma(p-1) > \sigma(p) > \tau(m)$; et l'on pose

$$\sigma(d_{j-1}+1)\sigma(d_{j-1}+2)\dots\sigma(m-1)\sigma(m)\sigma(m+1)\dots\sigma(p-1)\sigma(p)$$

égal à

$$\tau(d_{j-1}+1)\tau(d_{j-1}+2)\dots\tau(m-1)\sigma(m+1)\sigma(m+2)\dots\sigma(p)\tau(m)$$

et aller en (6) ;
(5) si $\tau(m-1) > \tau(m)$ et $\sigma(m+1)$,
alors $\sigma(m) \leftarrow \tau(m)$ si $d_{j-1} = m-1$; sinon, on note p le plus petit indice tel que $d_{j-1}+1 \le p \le m-1$ et $\tau(m) < \tau(p) < \tau(p+1) < \dots < \tau(m-1)$; et l'on pose

$$\sigma(d_{j-1}+1)\sigma(d_{j-1}+2)\dots\sigma(p-1)\sigma(p)\sigma(p+1)\dots\sigma(m-1)\sigma(m)$$

égal à

$$\tau(d_{j-1}+1)\tau(d_{j-1}+2)\dots\tau(p-1)\tau(m)\tau(p)\dots\tau(m-2)\tau(m-1) \ ;$$

(6) si $j = 1$, l'algorithme est terminé et on pose

$$S^{-1}w \leftarrow \sigma(1)\ \sigma(2) \dots \sigma(n) \ ;$$

sinon décrémenter j de 1 et aller en (2).

Conservons les mêmes notations que dans la description de l'algorithme pour S^{-1}, à la différence que nous écrivons $w = s(1)\, s(2) \dots s(n)$ au lieu de $s_1\, s_2 \dots s_n$. Désignons de plus par $c_1^i\, c_2^i \dots c_{k_i}^i$ la suite croissante des indices m tels que $s(m) \leq i$ $(i = 1, 2, \dots, x)$. Pour obtenir $\sigma = S^{-1}w$, on détermine successivement $\sigma_i' = S^{-1}\, s(c_1^i)\, s(c_2^i) \dots s(c_{k_i}^i)$ en utilisant l'algorithme précédent, pour $i = 1, 2, \dots, x$. Naturellement $\sigma_x' = \sigma$. Ce procédé est utilisé dans l'exemple suivant.

EXEMPLE.

$w =$	0	1	0	1	0	3	2	3	3	3	2	0	2	3	
$v_1 =$	0	1	0	1	0	.	.	.	.	.	.	0	.	.	
$\tau_1 = \sigma_1' =$	4	6	3	5	2	.	.	.	.	.	.	1	.	.	
$v_1 = v_1' =$	0	1	0	1	0	.	.	.	.	.	.	0	.	.	
$v_2 =$	0	1	0	1	0	.	2	.	.	.	2	0	2	.	
$\tau_2 =$	4	6	3	5	2	.	9	.	.	.	8	1	7	.	
$\sigma_2' =$	4	6	3	5	2	.	9	.	.	.	8	1	7	.	
$v_2' =$	0	1	0	1	0	.	2	.	.	.	2	0	2	.	
$v_3 =$	0	1	0	1	0	3	2	3	3	3	2	0	2	3	
$\tau_3 =$	9	11	3	10	2	8	14	7	6	5	13	1	12	4	
											13	1	4	12	(I5)
										13	5	1	4	12	(I4)
									13	6	5	1	4	12	(I4)
							7	14	13	6	5	1	4	12	(I5)
$\sigma_3' =$	9	11	3	10	2	8	7	14	13	6	5	1	4	12	(I3)

$\sigma = S^{-1}w = \sigma_3'$.

7. LA STATISTIQUE EULÉRIENNE IMAL

Rappelons que le Lehmer-code d'une permutation $\sigma = \sigma(1)\, \sigma(2) \dots \sigma(n)$ est la suite $L\sigma = x_1\, x_2 \dots x_3$ définie par

$$x_i = \operatorname{card} \{j : 1 \leq j \leq i-1 ,\ \sigma(j) < \sigma(i)\}$$

pour tout $i = 1, 2, \dots, n$.
Le code $L\sigma$ appartient à l'ensemble D_n. Par ailleurs, il est immédiat que $L : \mathfrak{S}_n \to D_n$ est bijectif, et que, d'autre part, on a

$$\text{RISE SET } L\sigma = \text{RISE SET}\sigma$$

pour tout σ dans $\mathfrak{S}_n$. Rappelons encore que pour tout $w = x_1\, x_2 \dots x_n$ dans D_n on note $\text{IMA}w$ le nombre de x_i distincts dans w et que l'on pose

IMAL = IMA $\circ$ L .

Nous avons vu dans la section précédente que la bijection $S : \mathfrak{S}_n \to D_n$ avait les propriétés suivantes

(i) RISE SET $S\sigma$ = RISE SETσ

(ii) IMA $S\sigma$ = IRISE σ .

En composant S et L^{-1} , on obtient

$$\mathfrak{S}_n \overset{S}{\to} D_n \overset{L^{-1}}{\to} \mathfrak{S}_n .$$

D'où

$$\begin{aligned} &\text{RISE SET } L^{-1} \circ S\sigma = \text{RISE SET}\,\sigma \\ &\text{IMAL } L^{-1} \circ S\sigma = \text{IMA } S\sigma = \text{IRISE}\,\sigma . \end{aligned} \tag{16}$$

Les deux permutations $L^{-1} \circ S\sigma$ et σ ayant même RISE SET, ont aussi même MAJ . Les identités (16) montrent alors que <u>les couples</u> (IRISE, MAJ) <u>et</u> (IMAL, MAJ) <u>ont même distribution sur</u> $\mathfrak{S}_n$. En fait les identités (16) contiennent le résultat plus fort suivant.

THÉORÈME 3. <u>Pour tout ensemble</u> T <u>tel que</u> $0 \in T \subset [n-1]$ <u>les statistiques</u> IRISE <u>et</u> IMAL <u>ont même distribution sur l'ensemble</u> $\{\sigma \in \mathfrak{S}_n : \text{RISE SET } \sigma = T\}$.

On trouvera une illustration de ce théorème dans la table 3 construite pour $n = 4$.

RISE SET	σ	Lσ	IMALσ	IRISE σ
0,1,2,3	1 2 3 4	0 1 2 3	4	4
0,1,2	1 2 4 3	0 1 2 2	3	3
	1 3 4 2	0 1 2 1	3	3
	2 3 4 1	0 1 2 0	3	3
0,1,3	1 3 2 4	0 1 1 3	3	3
	1 4 2 3	0 1 1 2	3	3
	2 3 1 4	0 1 0 3	3	3
	2 4 1 3	0 1 0 2	3	2
	3 4 1 2	0 1 0 1	2	3
0,2,3	2 1 3 4	0 0 2 3	3	3
	3 1 2 4	0 0 1 3	3	3
	4 1 2 3	0 0 1 2	3	3
0,1	1 4 3 2	0 1 1 1	2	2
	2 4 3 1	0 1 1 0	2	2
	3 4 2 1	0 1 0 0	2	2
0,2	2 1 4 3	0 0 2 2	2	2
	3 2 4 1	0 0 2 0	2	2
	3 1 4 2	0 0 2 1	3	3
	4 1 3 2	0 0 1 1	2	2
	4 2 3 1	0 0 1 0	2	2
0,3	3 2 1 4	0 0 0 3	2	2
	4 2 1 3	0 0 0 2	2	2
	4 3 1 2	0 0 0 1	2	2
0	4 3 2 1	0 0 0 0	1	1

Table 3 .

RÉFÉRENCES

L. Carlitz (1954), q-Bernoulli and Eulerian numbers, Trans. Amer. Math. Soc. **76** , 332-350 .

L. Carlitz (1956), The expansion of certain products, Proc. Amer. Math. Soc. **7** , 558-564 .

L. Carlitz (1959), Eulerian numbers and polynomials, Math. Magazine **33** , 247-260 .

L. Carlitz (1975), A combinatorial property of q-Eulerian numbers, Amer. Math. Monthly **82**, 51-54.

L. Carlitz, D. P. Roselle and R. A. Scoville (1966), Permutations and sequences with repetitions by number of increases, J. Combinatorial Theory **1**, 350-374.

M. S. Cheema and T. S. Motzkin (1971), Multipartitions and Multipermutations, Proc. of Symposia in Pure Math., vol. 19, Combinatorics, Amer. Math. Soc., Providence, 39-70.

L. Comtet (1970), Analyse Combinatoire, vol. 2, Presses Universitaires de France, Paris.

D. Dumont (1974), Interprétations combinatoires des nombres de Genocchi, Duke Math. J. **41**, 305-318.

W. Feller (1966), An Introduction to Probability Theory and its Applications, vol. 2, J. Wiley, New York, p. 27.

D. Foata (1968), On the Netto inversion number of a sequence, Proc. Amer. Math. Soc. **19**, 236-240.

D. Foata et M.-P. Schützenberger (1970), Théorie géométrique des polynômes eulériens, Lecture Notes in Math. n°138, Springer-Verlag, Berlin.

D. Foata et M.-P. Schützenberger (1977a), Major index and inversion number of permutations, A paraître dans Math. Nachrichten.

D. Foata et M.-P. Schützenberger (1977b), La troisième transformation fondamentale du groupe des permutations. A paraître.

Marquis de Laplace (1820), Oeuvres complètes, vol. 7, réédité par Gauthier-Villars (1886), Paris, p. 257 et suivantes.

R. von Randow und W. Meyer (1971), Ein Würfelschnittproblem und Bernoullische Zahlen, Math. Ann. **193**, 315-321.

J. Riordan (1958), An Introduction to Combinatorial Analysis, J. Wiley, New York.

D. P. Roselle (1974), Coefficients associated with the expansion of certain products, Proc. Amer. Math. Soc. **45**, 144-150.

R. P. Stanley (1976a), Binomial posets, Möbius inversion, and permutation enumeration, J. Combinatorial Theory series A, 20 , 336-356 .

R. P. Stanley (1976b), Eulerian partitions of a unit hypercube, voir note ci-après.

Eulerian Partitions of a Unit Hypercube

by Richard P. Stanley

In the preceding paper Dominique Foata mentions the result, implicit in the work of Laplace, that the volume of the region R_{nk} of the unit hypercube $[0,1]^n$ contained between the two hyperplanes $\Sigma x_i = k-1$ and $\Sigma x_i = k$ is given by $\frac{1}{n!} A_{nk}$, where A_{nk} is an Eulerian number. On the other hand, it follows from the well-known combinatorial interpretation of A_{nk} as the number of permutations of $\{1,2,\ldots,n\}$ with k rises (counting one rise at the start) that $\frac{1}{n!} A_{nk}$ is also the volume of the set S_{nk} of all points $(x_1,\ldots,x_n) \in [0,1]^n$ for which $x_i < x_{i+1}$ for exactly k values of i (including by convention $i = 0$). Foata raises the problem of whether there is some explicit measure-preserving map $\varphi : [0,1]^n \to [0,1]^n$ which takes S_{nk} onto R_{nk}, except possibly on a set of measure zero. We claim that such a map is given as follows : Define $\varphi : [0,1]^n \to [0,1]^n$ by $\varphi(x_1,\ldots,x_n) = (y_1,\ldots,y_n)$, where

$$y_i = \begin{cases} x_{i-1} - x_i & \text{if } x_i < x_{i-1} \\ 1 + x_{i-1} - x_i & \text{if } x_i > x_{i-1} \end{cases} .$$

Here we set $x_0 = 0$, and we leave φ undefined on the set of measure zero consisting of points where some $x_{i-1} = x_i$. If $(x_1,\ldots,x_n) \in S_{nk}$, then $\Sigma y_i = k-x_n$. Hence $(y_1,\ldots,y_n) \in R_{nk}$. Moreover, in each of the 2^{n-1} regions of $[0,1]^n$ determined by whether $x_i < x_{i-1}$ or $x_i > x_{i-1}$ for $2 \le i \le n$, φ is an affine transformation of determinant $(-1)^n$. Hence φ is measure-preserving. Finally, the inverse of φ is defined (except for the set of measure zero where some $y_1+y_2+\ldots+y_i$ is an integer) by $x_i = 1 + [y_1+y_2+\ldots+y_i] - y_1-y_2-\ldots-y_i$.

COHEN-MACAULAY COMPLEXES*

Richard P. Stanley

Department of Mathematics,
Massachusetts Institute of Technology,
Cambridge, Massachusetts

1. SIMPLICIAL COMPLEXES

Let Δ be a finite simplicial complex (or complex for short) on the vertex set $V = \{x_1,\ldots,x_n\}$. Thus, Δ is a collection of subsets of V satisfying the two conditions: (i) $\{x_i\} \in \Delta$ for all $x_i \in V$, and (ii) if $F \in \Delta$ and $G \subset F$, then $G \in \Delta$. There is a certain commutative ring A_Δ which is closely associated with the combinatorial and topological properties of Δ. We will discuss this association in the special case when A_Δ is a Cohen-Macaulay ring. Lack of space prevents us from giving most of the proofs and from commenting on a number of interesting sidelights. However, a greatly expanded version of this paper is being planned.

Let Δ be a complex (= finite simplicial complex). If $F \in \Delta$, we call F a face of Δ. If F has $i + 1$ elements (denoted card $F = i + 1$), we say dim $F = i$. Let $d = \delta + 1 = \max\{\text{card } F | F \in \Delta\}$. We write $\dim \Delta = \delta = d - 1$. If every maximal face of Δ has dimension δ, then Δ is called pure (or homogeneous by topologists). Let f_i be the number of i-dimensional faces of Δ. Thus $f_0 = n$. The vector $f = (f_0,f_1,\ldots,f_\delta)$ is called the f-vector of Δ. Now define a function on the non-negative integers by

$$H(\Delta,m) = \begin{cases} 1, & m = 0 \\ \sum_{i=0}^{\delta} f_i \binom{m-1}{i}, & m > 0. \end{cases} \tag{1}$$

Define integers h_i by

*Partially supported by NSF Grant # MCS 7308445-A04

M. Aigner (ed.), Higher Combinatorics, 51-62.

$$(1-\lambda)^d \sum_{m=0}^{\infty} H(\Delta,m)\lambda^m = \sum_{i=0}^{\infty} h_i\lambda^i .$$

It is easily seen that $h_i = 0$ if $i > d$. The vector $h=(h_0,h_1,\ldots,h_d)$ is called the h-vector of Δ. Knowing the f-vector of Δ is equivalent to knowing its h-vector.

Let $|\Delta|$ denote the underlying topological space of Δ, as defined in topology. The notation $\Delta = \langle abc, acd, bcd, bde\rangle$ means that the maximal faces of Δ are $\{a,b,c\}$, $\{a,c,d\}$, $\{b,c,d\}$, and $\{b,d,e\}$. For this Δ, the f-vector is (5,8,4), the h-vector is (1,2,1,0), and $|\Delta|$ is a 2-cell.

Let k be a field, fixed once and for all. All homology groups appearing in this paper are taken over the coefficient field k. Let $A = k[x_1,\ldots,x_n]$ be the polynomial ring over k whose variables are the vertices of Δ. Let I_Δ be the ideal of A generated by all squarefree monomials $x_{i_1}x_{i_2}\cdots x_{i_j}$ such that $\{x_{i_1},\ldots,x_{i_j}\} \not\in \Delta$. For instance, if $\Delta = \langle abc, acd, bcd, bde\rangle$, then $I_\Delta = (ae,ce,abd)$. (We only need to include the minimal "non-faces" of Δ as generators of I_Δ, since I_Δ is an ideal.) Let $A_\Delta = A/I_\Delta$. This ring was first considered by M. Hochster (who suggested it to his student G. Reisner [10] for further study) and independently by this writer [14] [15]. In order to study the algebraic properties of A_Δ, we will require some concepts from commutative algebra. We will survey these concepts in the context of "standard k-algebras," although much of what we say can be generalized considerably.

2. BETTI NUMBERS OF RINGS

Let $A = k[x_1,\ldots,x_n]$ as above, and let I be any homogeneous ideal of A (i.e., I is generated by homogeneous polynomials). Set R=A/I. We call such a ring R a standard k-algebra. In particular, R is graded as $R = R_0+R_1+\cdots$, where R_i is the k-vector space of homogeneous polynomials of degree i contained in R. Thus $R_0 = k$, R_1 generates R as a k-algebra, $R_iR_j \subset R_{i+j}$, and $\dim_k R_i < \infty$. The Hilbert function H(R,m) of R is defined by $H(R,m) = \dim_k R_m$. It was first shown by Hilbert that H(R,m) is a polynomial for m sufficiently large, the Hilbert polynomial of R. The Krull dimension of R, denoted dim R, can be defined to be one more than the degree of the Hilbert polynomial of R.

If R = A/I is a standard k-algebra, then a finite free resolution of R (as an A-module) is an exact sequence $0 \to M_j \to M_{j-1} \to \cdots \to M_0 \to R \to 0$ of A-modules, where each M_i is a free A-module of finite rank. A theorem of Hilbert implies that a finite free resolution of R always exists. There is a unique such resolution which minimizes the rank of each M_i; this resolution is called minimal. Define the i-th Betti number $\beta_i=\beta_i(R)$ of R to be the rank of the A-module M_i appearing in the minimal free resolution of R. In particular, $\beta_0=1$ and β_1 is the minimal number of generators of I. In the language of homological algebra, $\beta_i = \dim_k \mathrm{Tor}_i^A(R,k)$. For further information, see [12].

Example. Let $\Delta = \langle ab,bc,ac,cd\rangle$, so $I_\Delta = (ad,bd,abc)$. Then the minimal free resolution of A_Δ has the form $0 \to M_2 \to M_1 \to M_0 \to A_\Delta \to 0$, where rank $M_0 = 1$, rank $M_1 = 3$, rank $M_2 = 2$. With an appropriate choice of bases $\{X\}$ for M_0, $\{Y_1,Y_2,Y_3\}$ for M_1, and $\{Z_1,Z_2\}$ for M_2, the maps are given by $X \mapsto 1$, $Y_1 \mapsto adX$, $Y_2 \mapsto bdX$, $Y_3 \mapsto abcX$, $Z_1 \mapsto bY_1 - aY_2$, $Z_2 \mapsto bcY_1 - dY_3$. We have $\beta_0 = 1$, $\beta_1 = 3$, $\beta_2 = 2$, and $\beta_i = 0$ if $i \geq 3$.

If R is a standard k-algebra, let h be the largest integer i for which $\beta_i(R) \neq 0$. It is known that $n - d \leq h \leq n$, where $d = \dim R$ and n is the number of variables in A. The integer h is the *homological dimension* of R (as an A-module), denoted $\mathrm{hd}_A R$ or just hd R. If hd R = n - d then R is said to be a *Cohen-Macaulay ring*. The integer β_{n-d} is then called the *type* of R, denoted type R. If R is a Cohen-Macaulay ring of type one, then R is said to be *Gorenstein*. In this case, one can show $\beta_i = \beta_{h-i}$, where h = hd R. If R is Cohen-Macaulay, it is known that $\mathrm{Ext}_A^i(R,A) = 0$ unless i = hd R. Letting $\Omega(R) = \mathrm{Ext}_A^h(R,A)$, where h = hd R, this means that if we "dualize" the minimal free resolution of R by applying the functor $\mathrm{Hom}_A(\cdot, A)$, then we obtain a minimal free resolution for $\Omega(R)$, regarded as an A-module. $\Omega(R)$ is called the *canonical module* of R. Given that R is Cohen-Macaulay, one has that R is Gorenstein if and only if $\Omega(R) \cong R$. Thus the minimal free resolution of a Gorenstein standard k-algebra is "self-dual", a much stronger result than $\beta_i = \beta_{h-i}$.

3. CHARACTERIZING HILBERT FUNCTIONS

We now consider the relationship between the structure of R and its Hilbert function H(R,m). A non-void set M of monomials $x^\alpha y^\beta \cdots$ is called an *order ideal of monomials* if whenever $u \in M$ and v divides u, then $v \in M$. A finite or infinite sequence $h = (h_0,h_1,\ldots)$ of integers is called an *O-sequence* if there exists an order ideal M of monomials containing exactly h_i monomials of degree i. For instance, (1,3,2,2) is an O-sequence, since we can take $M = \{1, x, y, z, x^2, xy, x^3, x^2y\}$. A finite order ideal M of monomials is said to be *pure* if the maximal elements of M (ordered by divisibility) all have the same degree. We define a *pure O-sequence* in the obvious way. For instance, (1,3,1) is an O-sequence but not a pure O-sequence. Clearly, if $(h_0,h_1,\ldots)$ is an O-sequence, then $h_0 = 1$ and

$$0 \leq h_i \leq \binom{h_1 + i - 1}{i}, \tag{3}$$

since the corresponding order ideal M has h_1 variables actually appearing, and there are $\binom{h_1+i-1}{i}$ monomials of degree i in h_1 variables. An explicit numerical condition for a sequence $(h_0,h_1,\ldots)$ to be an O-sequence appears in [15], though the crux of this result was first proved by Macaulay. No similar characterization of pure O-sequences is known.

The next result characterizes the Hilbert function of a Cohen-Macaulay standard k-algebra. This result is due to Macaulay, but is first stated in "modern" terminology in [14]. See [16, Cor. 3.10] for a proof.

Theorem 1. Let H be a function on the non-negative integers, and let k be any field. Then H is the Hilbert function of a Cohen-Macaulay standard k-algebra of Krull dimension d if and only if the sequence $(h_0,h_1,\ldots)$ defined by

$$(1-\lambda)^d \sum_{m=0}^{\infty} H(m)\lambda^m = \sum_{i=0}^{\infty} h_i\lambda^i \tag{4}$$

is an O-sequence with finitely many non-zero terms.

If R is a Cohen-Macaulay standard k-algebra of Krull dimension d and Hilbert function H, then we call the sequence $h = (h_0,h_1,\ldots)$ defined by (4) the h-vector of R. If $h_i = 0$ for $i > s$, we also write $h = (h_0,h_1,\ldots,h_s)$ for this h-vector.

We are now in a position to define a concept intermediate between Cohen-Macaulay and Gorenstein which will be of interest to us. Suppose that R is a Cohen-Macaulay standard k-algebra with h-vector $(h_0,h_1,\ldots,h_s)$, $h_s \neq 0$. It is easy to see that $h_s \leq$ type R. If $h_s =$ type R, then we say that R is a level ring, and we call $(h_0,h_1,\ldots,h_s)$ a level sequence. A level ring with $h_s = 1$ is just a Gorenstein ring, and in this case we call $(h_0,h_1,\ldots,h_s)$ a Gorenstein sequence. Clearly, every level sequence is an O-sequence. Unlike the Cohen-Macaulay case, no characterization of level sequences, or even of Gorenstein sequences, is known. The next result gives some information about level sequences, though undoubtedly stronger restrictions can be obtained.

Theorem 2. Let $h = (h_0,h_1,\ldots,h_s)$ be a level sequence with $h_s \neq 0$.

(i) If i and j are non-negative integers with $i + j \leq s$, then $h_i \leq h_j h_{i+j}$. In particular, if h is a Gorenstein sequence then $h_i = h_{s-i}$.

(ii) The vector $(h_s,h_{s-1},\ldots,h_0)$ is a sum of h_s O-sequences.

(iii) If $0 \leq t \leq s$, then $(h_0,h_1,\ldots,h_t)$ is a level sequence.

For instance, (1,4,10,2) is an O-sequence but by Theorem 2(i) is not a level sequence. Similarly, (1,4,2,2) is an O-sequence but by Theorem 2(ii) is not a level sequence. On the other hand, (1,3,5,4,5,3,1) is an O-sequence but not a Gorenstein sequence, although this example is not covered by Theorem 2. A characterization of Gorenstein sequences with $h_1 \leq 3$ appears in [16, Thm. 4.2]. Finally, we remark that it is easily seen that every pure 0-sequence is a level sequence but not conversely, e.g., (1,3,1).

4. APPLICATIONS TO SIMPLICIAL COMPLEXES

We are now in a position to apply the above results on standard k-algebras to those of the form A_Δ. We begin with a simple result whose proof appears in [15].

Theorem 3. Let Δ be a complex with $d = 1 + \dim \Delta$. Then $d = \dim A_\Delta$, and the Hilbert function $H(A_\Delta, m)$ is the function $H(\Delta, m)$ of (1).

Corollary. Suppose A_Δ is Cohen-Macaulay. Then the h-vector of Δ is equal to the h-vector of A_Δ. Consequently, the h-vector of Δ is an 0-sequence.

The above corollary raises the question of determining for which Δ is A_Δ Cohen-Macaulay, or more generally of computing hd A_Δ. The answer to this question follows from the following unpublished result of M. Hochster. First we require some notation. Let V be the set of vertices of Δ, and let $W \subset V$. Let Δ_W denote the restriction of Δ to W, i.e., $\Delta_W = \{F \in \Delta | F \subset W\}$. Throughout this paper, the notation H (respectively, $\tilde{H}$) denotes homology (respectively, reduced homology), either simplicial or singular (whichever is appropriate), over the coefficient field k, with the conventions $\tilde{H}_{-1}(\Gamma) = 0$ unless $\Gamma = \phi$, $\tilde{H}_i(\phi) = 0$ if $i \geq 0$, $\tilde{H}_{-1}(\phi) = k$, $\tilde{H}_i(\Gamma) = 0$ if $i < -1$.

Theorem 4. The Betti numbers of A_Δ are given by

$$\beta_i(A_\Delta) = \sum \dim_k \tilde{H}_{j-i-1}(\Delta_W) ,$$

where the sum is over all subsets W of the set V of vertices of Δ, and where card $W = j$.

Theorem 4 yields a topological criterion for computing hd A_Δ and therefore determining whether or not A_Δ is Cohen-Macaulay, but this criterion is quite cumbersome to use. A simpler condition for A_Δ to be Cohen-Macaulay was given by G. Reisner [10] prior to the discovery of Theorem 4. The equivalence of (i) and (ii) below is Reisner's result, while the equivalence of (ii) and (iii) is a simple exercise in topology. First recall that if $F \in \Delta$, then the link of F is defined by

$$\ell k\ F = \{G \in \Delta | F \cap G = \phi \text{ and } F \cup G \in \Delta\}.$$

In particular, $\ell k\ \phi = \Delta$.

Theorem 5. The following three conditions are equivalent.

(i) A_Δ is Cohen-Macaulay.

(ii) For all $F \in \Delta$ (including $F = \phi$), $\tilde{H}_i(\ell k\ F) = 0$ if $i \neq \dim \ell k\ F$.

(iii) If $X = |\Delta|$, then $\tilde{H}_i(X) = H_i(X, X-p) = 0$ for all $p \in X$ and $i \neq \dim X$.

When A_Δ is Cohen-Macaulay, we call Δ a <u>Cohen-Macaulay complex</u>. The property of being a Cohen-Macaulay complex depends on k. It follows, however, from the Universal Coefficient Theorem that Δ is Cohen-Macaulay over <u>some</u> k if and only if it is Cohen-Macaulay over the rational numbers.

Note that Theorem 5 implies that the question of whether or not Δ is Cohen-Macaulay depends only on $|\Delta|$ (and the coefficient field k). Recently J. Munkres has shown, using Theorem 4, that for <u>any</u> Δ the integer n-hd A_Δ depends only on $|\Delta|$ (and k).

If $h = (h_0, h_1, \ldots, h_d)$ is the h-vector of a Cohen-Macaulay complex Δ, then by the corollary to Theorem 3 and (3), $h_i \leq \binom{n-d+i-1}{i}$. When $|\Delta|$ is a sphere, this condition is equivalent to a condition on the f-vector of Δ of the form $f_i \leq c_i(n,d)$, where $c_i(n,d)$ is a certain explicit number depending on H on i,n, and d. A complex satisfying the above condition on h_i is said to satisfy the <u>Upper Bound Conjecture</u> (UBC). If $X = |\Delta|$ is a topological manifold with or without boundary, then $H_i(X, X-p) = 0$ for all $p \in X$ and $i < \dim X$. Hence by Theorem 5, Δ is Cohen-Macaulay if and only if $\tilde{H}_i(X) = 0$ for $i < \dim X$. In particular, Δ is Cohen-Macaulay if Δ is a sphere or cell. Thus the UBC holds for spheres and cells. For further details, see [15]. An example of a complex which fails to satisfy the UBC is <abcd,ae,be,ce>, whose h-vector is (1,1,0,-3,2). No example is known of a complex Δ which fails to satisfy the UBC for which $|\Delta|$ is a manifold with or without boundary.

5. CONSTRUCTIBILITY AND SHELLABILITY

We now give a result which shows that the corollary to Theorem 3 completely characterizes the h-vector of a Cohen-Macaulay complex. We say that a complex Δ is <u>constructible</u> (a concept due to M. Hochster) if it can be obtained by the following recursive procedure: (i) any simplex is constructible, and (ii) if Δ' and Δ'' are constructible of the same dimension δ, and if $\Delta' \cap \Delta''$ is constructible of dimension $\delta-1$, then $\Delta' \cup \Delta''$ is constructible. A straightforward Mayer-Vietoris argument, combined with Theorem 5, shows that Δ is Cohen-Macaulay if it is constructible. (A simple direct algebraic proof can also be given; see [14].)

Suppose that in building up a constructible polytope, one can always take Δ'' to be a simplex. Equivalently, Δ is pure and its maximal faces can be ordered $F_1, F_2, \ldots, F_\mu$ so that for $i = 2, 3, \ldots, \mu$, we have that $(F_1 \cup F_2 \cup \cdots \cup F_{i-1}) \cap F_i$ is a non-void union of faces F of F_i satisfying $\dim F_i - \dim F = 1$. Then Δ is said to be <u>shellable</u>. This differs somewhat from other definitions of "shellable" in the literature, in that we place no restrictions on when $(F_1 \cup \cdots \cup F_{i-1}) \cap F_i$ can consist of <u>all</u>

faces F of F_i of codimension one. See [5] for an interesting account of shellable complexes.

Theorem 6. Let $h = (h_0, h_1, \ldots, h_d)$ be a sequence of integers. The following four conditions are equivalent.

(i) h is an 0-sequence,

(ii) h is the h-vector of a Cohen-Macaulay complex Δ,

(iii) h is the h-vector of a constructible complex Δ,

(iv) h is the h-vector of a shellable complex Δ.

The most important examples of shellable complexes are the boundary complexes of simplicial convex polytopes. We do not know an example of a constructible complex which is not shellable. However, in Section 7 are examples of constructible complexes for which it is unclear whether they are shellable; and it seems quite likely that a constructible complex need not be shellable.

6. GORENSTEIN COMPLEXES

If A_Δ is Gorenstein then Δ is called a Gorenstein complex. (As usual, this depends on k). We now give a characterization of Gorenstein complexes which can be deduced from Theorem 4 by using either topological or combinatorial arguments. Recall that if Γ and Δ are complexes on disjoint vertex sets V and W, then their join $\Gamma * \Delta$ is a complex on $V \cup W$ defined by $\Gamma * \Delta = \{F \cup G \mid F \in \Gamma \text{ and } G \in \Delta\}$.

Theorem 7. Δ is a Gorenstein complex if and only if it is a join $\sigma * \Gamma$, where σ is a simplex and where

$$\tilde{H}_i(X) = 0 \text{ for } i < \delta; \ \dim_k \tilde{H}_\delta(X) = 1$$

$$H_i(X, X-p) = 0 \text{ for } i < \delta; \ \dim_k H_\delta(X, X-p) = 1 \text{ for all } p \in X, \quad (5)$$

where $X = |\Gamma|$ and $\delta = \dim X$.

Since (5) automatically holds when X is a manifold, we see in particular that Δ is Gorenstein if X is a sphere, a result first proved by M. Hochster (unpublished). We also remark that it is possible for Δ to be Gorenstein (over any field k) without X being a topological manifold, e.g., when X is the suspension of Kneser's "dodecahedral space."

Suppose $(h_0, h_1, \ldots, h_d)$ is the h-vector of a Gorenstein complex Δ. We may assume that $h_d \neq 0$, since if $h_s \neq 0$ and $h_{s+1} = 0$, then $(h_0, \ldots, h_s)$ is the h-vector of the complex Γ of Theorem 7. By Theorem 2(ii), we then have $h_i = h_{d-i}$. This relation is equivalent to a condition on the f-vector of Δ known as the Dehn-Sommerville equations.

An outstanding open problem in the theory of convex polytopes is to characterize the h-vector $(h_0, \ldots, h_d)$ of the boundary complex of a simplicial convex d-polytope. McMullen's still open "g-conjecture" [9] states that the desired characterization is given by the following two conditions:

$$h_i = h_{d-i} \text{ for all } i,$$
$$(h_0, h_1 - h_0, h_2 - h_1, \ldots, h_e - h_{e-1}) \text{ is an 0-sequence, where } e = [d/2]. \quad (6)$$

We ask whether (6) also holds when $|\Delta|$ is a sphere or even more generally when Δ is Gorenstein. There are special cases for which it is possible to verify (6). For instance, in [9] it is shown by geometric means (Gale diagrams) that if $|\Delta|$ is a sphere satisfying $n \leq 4 + \dim \Delta$, then (6) holds. This is also an immediate consequence of Theorem 7 and [16, Thm. 4.2], and in fact one needs only to assume that Δ is a Gorenstein complex satisfying $n \leq 4 + \dim \Delta$. The next theorem gives another such result. It is an easy consequence of Theorem 4 and the corollary to Theorem 3. First we require a definition. If $|\Delta|$ is a δ-dimensional manifold with boundary, then the boundary complex $\partial\Delta$ of Δ is the complex whose maximal faces are the $(\delta-1)$-dimensional faces of Δ which lie on only one δ-dimensional face. Thus $|\partial\Delta| = \partial|\Delta|$, so if $|\Delta|$ is a δ-dimensional cell, then $|\partial\Delta|$ is a $(\delta-1)$-dimensional sphere.

Theorem 8. Suppose that $|\Delta|$ is a d-dimensional manifold with boundary such that Δ is Cohen-Macaulay and $\partial\Delta$ is Gorenstein (e.g., $|\Delta|$ is a cell), and such that any face $F \in \Delta - \partial\Delta$ satisfies $\dim F \geq \frac{1}{2}(d-1)$. Then the h-vector $(h_0, h_1, \ldots, h_d)$ of $\partial\Delta$ satisfies (6).

A result of Klee [8] implies that if Δ is Gorenstein with h-vector $(h_0, \ldots, h_d)$, $h_d \neq 0$, then $h_1 + h_2 + \cdots + h_{d-1} \geq (d-1)h_1$. In [16] it is shown that (1,13,12,13,1) is a Gorenstein sequence. It follows that a Gorenstein sequence need not be the h-vector of a Gorenstein complex, in contrast to the Cohen-Macaulay case.

As a generalization of Theorem 7, we can ask for a description of the canonical module $\Omega(A_\Delta)$ when A_Δ is Cohen-Macaulay. If $|\Delta|$ is a manifold with boundary, there is overwhelming evidence (but not yet a proof) that $\Omega(A_\Delta)$ is isomorphic to the ideal of A_Δ generated by the squarefree monomials $x_{i_1} x_{i_2} \cdots x_{i_j}$ for which $\{x_{i_1}, \ldots, x_{i_j}\} \in \Delta - \partial\Delta$.

7. INDEPENDENT SETS AND BROKEN CIRCUITS

We now discuss some applications of Cohen-Macaulay complexes to the theory of pregeometries (or "matroids") in the sense of Crapo-Rota [4]. A finite pregeometry Γ consists of a finite set V of vertices (or "points"), and a collection Δ of subsets of V, called independent sets, such that (i) Δ is a complex, and (ii) for every subset W of V, the induced complex Δ_W is pure. To avoid trivialities, we will also assume $\{v\} \in \Delta$ for all $v \in V$, so we may identify Γ with Δ. We call Δ a G-complex. For example, if V is a finite set of points in a vector space and Δ is the collection of linearly independent subsets of V, then Δ is a G-complex. If V is the set of edges of a finite graph G and Δ is the collection of subsets of V containing no cycle, then Δ is a G-complex. For further examples, see [4]. We also refer the reader to [4] for any unexplained terminology in this section.

It is easy to see, using the so-called "Tutte-Grothendieck decomposition" [2], that a G-complex Δ is constructible and is therefore Cohen-Macaulay. Using Theorem 4, one can obtain a simple expression for the Betti numbers $\beta_i(A_\Delta)$. We need to compute $N = \dim_k \tilde{H}_{j-i-1}(\Delta_W)$ where $W \subset V$ and card $W = j$. Let $\delta = \dim \Delta_W$. Now Δ_W is a Cohen-Macaulay complex, so $N = 0$ unless $j-i-1 = \delta$. If $j-i-1 = \delta$, then $N = (-1)^\delta(\chi(\Delta_W)-1)$, where χ is the Euler characteristic. It is known that $\chi(\Delta_W) - 1 = 0$ unless V-W is a flat (closed set) of the dual pregeometry $\tilde{\Delta}$. When V-W is a flat, then $N = |\tilde{\mu}(V-W,V)|$, where $\tilde{\mu}$ is the Möbius function (in the sense of [11]) of the lattice $L(\tilde{\Delta})$ of flats of $\tilde{\Delta}$. Moreover, (card W)-δ-1 is just the corank of V-W in $\tilde{\Delta}$, i.e., the length of the longest chain between V-W and V in $L(\tilde{\Delta})$. Hence we obtain:

<u>Theorem 9</u>. Let Δ be a G-complex on a vertex set V. Then $\beta_i(A_\Delta) = \Sigma|\tilde{\mu}(X,V)|$, where X ranges over all flats of $\tilde{\Delta}$ of corank i.

Compare this with the so-called "Whitney number of the first kind" $\Sigma|\tilde{\mu}(\phi,X)|$, where X ranges over all flats of $\tilde{\Delta}$ of rank i. Theorem 9 implies that when Δ is a G-complex, the type of A_Δ is $|\tilde{\mu}(\phi,V)|$. On the other hand, if $(h_0,h_1,\ldots,h_d)$ is the h-vector of <u>any</u> complex Δ satisfying $\dim \Delta = d-1$, then an easy computation reveals $h_d = (-1)^{d-1}(\chi(\Delta)-1)$. Hence if Δ is a G-complex then $h_d = |\tilde{\mu}(\phi,V)| = \text{type } A_\Delta$. There follows:

<u>Corollary</u>. If Δ is a G-complex, then A_Δ is a level ring of type $|\tilde{\mu}(\phi,V)|$. Hence the h-vector of Δ is a level sequence.

Not every level sequence is the h-vector of some G-complex, e.g., (1,3,1), and it would be of considerable interest to characterize such h-vectors. In this direction, we have:

<u>Conjecture</u>. If Δ is a G-complex, then the h-vector of Δ is a pure O-sequence (as defined in Section 5).

Closely related to G-complexes are the "broken circuit complexes." Let $x_1,x_2,\ldots,x_n$ be an ordering of the vertices of a pregeometry Δ. A <u>broken circuit</u> is obtained by deleting the highest labeled element from any circuit (= minimal dependent set) of Δ. The <u>broken circuit complex</u> (or <u>BC-complex</u>) of Δ with respect to the ordering $x_1,\ldots,x_n$ is the complex whose faces are the subsets of V which do not contain a broken circuit. Let Λ denote the broken circuit complex of Δ (with respect to the ordering $x_1,\ldots,x_n$). If $(f_0,f_1,\ldots,f_\delta)$ is the f-vector of Λ, where $\delta = \dim \Lambda$, then $p_\Delta(\lambda) = \lambda^{\delta+1} - f_0\lambda^\delta + \cdots - (-1)^\delta f_\delta$ is the <u>characteristic polynomial</u> of Δ and is thus independent of the ordering chosen for the vertices (see [11, §7]). If Δ consists of the acyclic sets of edges of a graph G, then $\lambda^c \cdot p_\Delta(\lambda)$ is the <u>chromatic polynomial</u> of G, where c is the number of components of G. Hence the theory of Cohen-Macaulay complexes is applicable to chromatic polynomials.

It is easy to see that a BC-complex Λ is constructible and therefore Cohen-Macaulay. Hence the h-vector of Λ is an O-sequence. This improves an observation of Wilf [17, Thm. 2], who was the first person to study broken circuit complexes _qua_ complexes. For additional information on BC-complexes, see [3].

The fact that the h-vector of a BC-complex Λ is an O-sequence by no means characterizes such h-vectors, and it would be extremely interesting to obtain additional restrictions by a more detailed analysis of A_Λ. It has been conjectured that the f-vector $(f_0,\ldots,f_\delta)$ of a BC-complex is _unimodal_, i.e., for some i we have $f_0 \leq f_1 \leq \cdots \leq f_i$, $f_i \geq f_{i+1} \geq \cdots \geq f_\delta$. Unfortunately, this fact does not follow simply from the h-vector being an O-sequence. For instance, the vector h = (1, 500, 55, 220, 715, 2002) is an O-sequence, and the corresponding f-vector (with d = 5) is f = (1, 505, 2065, 3395, 3325, 3493). Hence, by Theorem 6, f is the f-vector of some Cohen-Macaulay (or even shellable) complex Δ. We remark that without Theorem 6 it is difficult to find an example even of a pure complex whose f-vector is not unimodal.

8. COHEN-MACAULAY POSETS

Let P be a finite poset (= partially ordered set). Let $\Delta(P)$ denote the complex whose vertices are the elements of P and whose faces are the chains (totally ordered subsets) of P. If $\Delta(P)$ is a Cohen-Macaulay complex, then we call P a _Cohen-Macaulay poset_. (As usual, this depends on k.) There are two main classes of such posets known. (i) A finite semimodular lattice is a Cohen-Macaulay poset. This follows from Theorem 5 and work of Folkman [7] and Farmer [6]. More generally, we conjecture that a finite _admissible lattice_ in the sense of [13] is Cohen-Macaulay. (ii) Let Σ be a finite regular cell complex, e.g., a finite simplicial complex. (Certain more general structures can be allowed.) Suppose that the underlying topological space of Σ satisfies condition (iii) of Theorem 5. If P is the set of faces of Σ, ordered by inclusion, then P is Cohen-Macaulay. Indeed, $\Delta(P)$ is just the first barycentric subdivison of Σ.

Cohen-Macaulay posets were first considered explicitly by Baclawski [1]. His Theorems 6.1 and 6.2 are special cases of the next result, which was conjectured by this writer and proved (unpublished) by J. Munkres. First, let us define the _rank_ $\rho(x)$ of an element x of a finite poset P to be the length of the longest chain of P whose top element is x. Thus $\rho(x) = 0$ if and only if x is a minimal element of P. If $x \leq y$ in P, set $\rho(x,y) = \rho(y) - \rho(x)$. If $\Delta(P)$ is pure, then $\rho(x,y)$ is the length of _any_ unrefinable chain between x and y, and $\rho(x,y) = \dim \Delta(P)$ if and only if x is a minimal element and y is a maximal element of P.

Theorem 10. Let P be a Cohen-Macaulay poset with rank function ρ, and let i be a non-negative integer. Let P_i be the set of all $x \in P$ satisfying $\rho(x) \neq i$. Give P_i the ordering induced from P. Then P_i is a Cohen-Macaulay poset.

If P is a Cohen-Macaulay poset with Möbius function μ, and if $x \leq y$ in P, then the open interval (x,y) is a Cohen-Macaulay poset and $(-1)^{\ell}\mu(x,y) = \dim_k \tilde{H}_{\ell}(\Delta((x,y)))$, where $\ell = \rho(x,y)$. Hence, $(-1)^{\ell}\mu(x,y) \geq 0$, i.e., the Möbius function of P alternates in sign. Theorem 10 implies that if we remove any set of "levels" from P, the Möbius function of the resulting poset continues to alternate in sign.

If P is a finite poset for which $\Delta(P)$ is pure, and if $\mu(x,y) = (-1)^{\rho(x,y)}$ for every $x \leq y$ in P, then P is called an Eulerian poset. It is not hard to see that a Cohen-Macaulay poset P is Gorenstein if and only if when we remove from P all elements x which are related to every element of P, and then adjoin a unique maximal and unique minimal element, the resulting poset is Eulerian.

The above considerations suggest that the Cohen-Macaulay posets are a natural class of posets whose Möbius functions merit further study.

REFERENCES

1. K. Baclawski, Homology and Combinatorics of Ordered Sets, thesis, Harvard University, 1976.
2. T. Brylawski, A decomposition for combinatorial geometries, Trans. Amer. Math. Soc. 171 (1972), 235-282.
3. T. Brylawski, The broken circuit complex, to appear.
4. H. Crapo and G.-C. Rota, On the Foundations of Combinatorial Theory: Combinatorial Geometries, M.I.T. Press, Cambridge, Massachusetts, 1970.
5. G. Danaraj and V. Klee, Which spheres are shellable?, preprint.
6. F. Farmer, Cellular homology for posets, preprint.
7. J. Folkman, The homology groups of a lattice, J. Math. Mech. 15 (1966), 631-636.
8. V. Klee, A d-pseudomanifold with f_0 vertices has at least $df_0 - (d-1)(d+2)$ d-simplices, Houston J. Math. 1 (1975), 81-86.
9. P. McMullen, The number of faces of simplicial polytopes, Israel J. Math. 9 (1971), 559-570.
10. G. Reisner, Cohen-Macaulay quotients of polynomial rings, Advances in Math. 21 (1976), 30-49.
11. G.-C. Rota, On the foundations of combinatorial theory, I. Theory of Möbius functions, Z. Wahrscheinlichkeitstheorie 2 (1964), 340-368.

12. J.-P. Serre, Algèbre Locale-Multiplicités, Lecture Notes in Math., no. 11, Springer-Verlag, Berlin, 1965.
13. R. Stanley, Finite lattices and Jordan-Hölder sets, Algebra Universalis 4 (1974), 361-371.
14. R. Stanley, Cohen-Macaulay rings and constructible polytopes, Bull. Amer. Math. Soc. 81 (1975), 133-135.
15. R. Stanley, The Upper Bound Conjecture and Cohen-Macaulay rings, Studies in Applied Math. 54 (1975), 135-142.
16. R. Stanley, Hilbert functions of graded algebras, Advances in Math., to appear.
17. H.S. Wilf, Which polynomials are chromatic?, Proc. Colloquium on Combinatorial Theory, Rome, 1973, to appear.

Late note. The work of M. Hochster to which we have referred (especially our Theorem 4) appears in Hochster's paper "Cohen-Macaulay rings, combinatorics, and simplicial complexes", based on a talk presented at the Oklahoma Ring Theory Conference, March 11-13, 1976. This paper contains many other interesting results on the structure of the ring A_Δ.

Later note. Regarding the conjecture in Section 6 concerning $\Omega(A_\Delta)$ when $|\Delta|$ is a manifold with boundary, Hochster has proved the following result. Suppose Δ is Cohen-Macaulay and $|\Delta|$ is a manifold with boundary. Let I be the ideal of A_Δ generated by all square free monomials $x_{i_1}x_{i_2}\cdots x_{i_j}$ for which $\{x_{i_1},\ldots,x_{i_j}\} \in \Delta - \partial\Delta$. Then I is isomorphic to $\Omega(A_\Delta)$ if and only if $\partial\Delta$ is Gorenstein (e.g., if $|\Delta|$ is orientable).

PART II

COMBINATORIAL SET THEORY AND ORDER THEORY

ACYCLIC ORIENTATIONS

(Notes from the talk of C. Greene)

1. ARRANGEMENTS OF HYPERPLANES

Let G be a finite undirected graph with vertex set V and edge set E. An acyclic orientation of G is an orientation of the edges of G such that there are no directed cycles. If L(G) denotes the bond lattice of G, μ the Möbius function of L, and $N(\pi)$ the number of components of $\pi \in L(G)$, then a well-known formula (see e.g. [3]) states that the chromatic polynomial $p_G(\lambda)$ of G can be expressed in the form

$$p_G(\lambda) = \sum_{\pi \in L(G)} \mu(0,\pi)\lambda^{N(\pi)}$$

<u>Theorem (Stanley [4])</u>

$$\#\text{ acyclic orientations of } G = |p_G(-1)| = \sum_{\pi \in L(G)} |\mu(0,\pi)|$$

The following considerations generalize Stanley's theorem and show that many combinatorial invariants associated with G can be interpreted as counting acyclic orientations of G with certain restrictions.

Let $\mathscr{H}$ be a finite set of hyperplanes in $\mathbb{R}^d$ and $\mathscr{L}(\mathscr{H})$ the lattice of nonempty intersections of $\mathscr{H}$ ordered by reverse inclusion. $p_{\mathscr{H}}(\lambda) = \sum_{A \in \mathscr{L}(\mathscr{H})} \mu(0,A)\lambda^{d(A)}$ is the characteristic polynomial of $\mathscr{L}(\mathscr{H})$.

M. Aigner (ed.), Higher Combinatorics, 65-68.

Theorem (Zaslavsky [5])

$$\# \text{ regions into which } \mathbb{R}^d \text{ is partitioned by } \mathcal{H} = |p_{\mathcal{H}}(-1)| = \sum_{A \in \mathcal{L}(\mathcal{H})} |\mu(0,A)|.$$

In the special case that the hyperplanes of $\mathcal{H}$ are in general position, each interval of $\mathcal{L}(\mathcal{H})$ is a Boolean Algebra and we obtain:

Corollary (Buck [1])

$$\# \text{ regions induced by } N \text{ hyperplanes of } \mathbb{R}^d \text{ in general position} = \binom{N}{0} + \binom{N}{1} + \dots + \binom{N}{d}.$$

To obtain Stanley's theorem associate to each edge $e = (a,b)$ of a graph $G(V,E)$ the hyperplane $H_e : x_a = x_b$ in $\mathbb{R}^N$, $N = |V|$. For the arrangement $\mathcal{H}(G) = \{H_e : e \in E\}$ we have $\mathcal{L}(\mathcal{H}(G)) \cong L(G)$. If we let the orientation $a \to b$ in G correspond to $x_a < x_b$ in $\mathcal{H}$ we get a bijection between acyclic orientation of G and regions induced by $\mathcal{H}$.

2. REFINEMENTS AND APPLICATIONS

Bounded regions theorem (Zaslavsky [5])

$$\# \text{ bounded regions in } \mathbb{R}^d \text{ induced by } \mathcal{H} = \Big|\sum_{A \in \mathcal{L}(\mathcal{H})} \mu(0,A)\Big|.$$

Suppose $\mathcal{H} = \{H_i : i \in I\}$ with $\bigcap_{i \in I} H_i = \{0\}$. Then it is obvious (and follows also from the bounded regions theorem) that there are no bounded regions induced by $\mathcal{H}$. Choose $H_0 \in \mathcal{H}$ and perturb H_0 away from the origin, i.e. replace the defining equation $h(x) = 0$ by $h(x) = \varepsilon$. Then there will be bounded regions in the resulting arrangement $\mathcal{H}_0'$.

Theorem

$$\# \text{ bounded regions in } \mathbb{R}^d \text{ induced by } \mathcal{H}_0' = \Big| \frac{dp_{\mathcal{H}}'(\lambda)}{d\lambda} \lambda = 1 \Big|$$

$$= \Big|\sum_{A \in \mathcal{L}(\mathcal{H})} d(A)\mu(0,A)\Big| = \beta(\mathcal{L}) \qquad (\beta\text{-Invariante})$$

$$= \Big|\sum_{\substack{A \in \mathcal{L} \\ A \not\leq H}} \mu(0,A)\Big| \text{ for any } H \in \mathcal{H}.$$

In particular, the number of bounded regions is independent of the hyperplane that was moved from the origin.

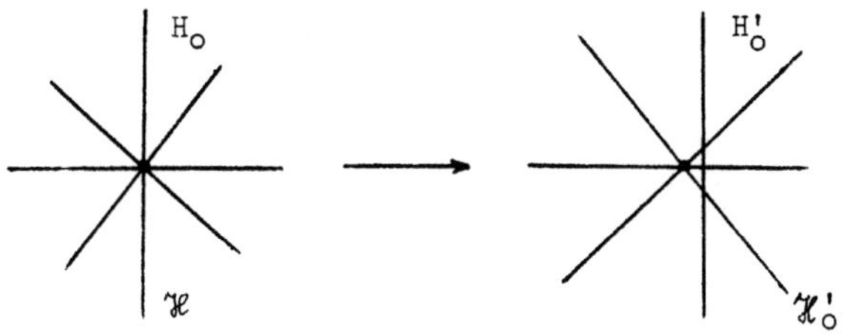

Corollary

a. # regions depends only on $\mathcal{L}(\mathcal{H})$.
b. # bounded regions depends only on $\mathcal{L}(\mathcal{H})$.
c. # regions incident to a certain hyperplane $H \in \mathcal{H}$ is independent of H.

For a graph $G(V,E)$, $a \in V$, $e = (b,c) \in E$ let $N(a) = \#$ acyclic orientations such that a is the unique source, and $N(e) = \#$ acyclic orientations such that b is the unique source and c is the unique sink.

Theorem

$N(a) = |\mu_{L(G)}(0,1)|, \quad N(e) = \beta_{L(G)} = |p'_G(1)|$

N(a) and N(e) are thus independent of a and e, respectively.

Examples

For the complete graph K_n we have

$L(K_n) = \mathcal{P}(n)$ (partition lattice), $|\mu| = (n-1)!$, $\beta = (n-2)!$.

For a tree T on n vertices we obtain

$p_T(\lambda) = \lambda(\lambda-1)^{N-1}$, $|\mu| = 1$, $\beta = 0$.

In general, $\beta = 0 \Leftrightarrow G$ is a separable graph, which is a well-known result for graphic matroids (see [2]).

3. TOTALLY CYCLIC ORIENTATIONS

The foregoing considerations can be generalized to a class of matroids $\underline{G}(S)$ where every circuit has a positive and a negative part (extending ± in vector spaces). A cycle is a positive circuit and $\underline{G}(S)$ is called acyclic if it contains no cycles. Zaslavsky's theorem generalizes to this situation.

Let us consider a real matrix H and the matroid $\underline{H}$ on its columns. Then

$$\underline{H} \text{ acyclic} \Leftrightarrow \nexists y \text{ with } Hy = 0,\ y \text{ semipositive},$$

and it follows easily for the dual matroid $\underline{H}^*$

$\underline{H}$ acyclic $\Leftrightarrow$ $\underline{H}^*$ totally cyclic (i.e. every column has the property that there is a nonnegative linear dependence involving it).

For a directed graph G to be totally cyclic means that every edge is contained in a consistently oriented circuit.

Theorem

$$\#\text{ totally cyclic orientations of } G = |p_{G^*}(-1)|.$$

REFERENCES

1. Buck R.C., Partition of space. Amer.Math.Monthly 50(1943), 541-544.
2. Crapo H.H., A higher invariant for matroids. J.Comb.Theory 2 (1967), 406-417.
3. Rota G.C., On the foundations of combinatorial theory I: Theory of Möbius functions. Z. Wahrscheinlichkeitsrechnung 2 (1964), 340-368.
4. Stanley R., Acyclic orientations of graphs. Discrete Math. 5 (1973), 171-178.
5. Zaslavsky T., Facing up to arrangements: face count formulas for partitions of space by hyperplanes. Mem. Amer. Math. Soc. 1(1975), issue 1, no. 154.

PROBLEMS IN COMBINATORIAL SET THEORY

Richard Rado

Department of Mathematics, The University,
Reading, England

This paper deals with two branches of combinatorial set theory, the partition calculus and the theory of Δ-systems. It presents a survey of some published work as well as some new results.

NOTATION

Roman capital letters denote sets. If nothing is said to the contrary, small Roman letters denote cardinal numbers; c^+ denotes the least cardinal greater than c, and we put $c^- = \min\{x : x^+ \geqslant c\}$. The infinite cardinals are $\aleph_0, \aleph_1, ..$ and the corresponding ordinals are $\omega_0 (=\omega), \omega_1, ...$ If $a \geqslant \aleph_0$ then cf a denotes the least c such that there is a representation $a = \sum(i \in I) a_i$ with $|I| = c$ and $a_i < a$ for $i \in I$. The cardinal of A is $|A|$ and we put

$$[A]^r = \{X \subseteq A : |X| = r\}.$$

The symbol $\{x_i : i \in I\}_{\neq}$ denotes the set $\{x_i : i \in I\}$ and at the same time expresses the condition that $x_i \neq x_j$ for $i \neq j$. Occasionally we use the obliteration operator ^ whose effect is to delete from a sequence the term above which it is placed. Whenever a set S_i has been defined for each $i \in I$ then we put $S_J = \bigcup(i \in J) S_i$ for $J \subseteq I$.

PART I. PARTITION RELATIONS

(a) The well known box argument states that if f is a function, $f : A \to I$, and if $|A| > |I|$, then there is $\{x,y\}_{\neq} \subseteq A$ with $f(x) = f(y)$. F.P. Ramsey [1] discovered the following remarkable extension of this principle.

M. Aigner (ed.), Higher Combinatorics, 69-78. All Rights Reserved.

Theorem 1.

If $|A| = \aleph_0$; $r, |I| < \aleph_0$; $f : [A]^r \to I$, then there is $X \in [A]^{\aleph_0}$ such that f is constant on $[X]^r$.

This theorem has given rise to the creation of the partition calculus. See, for instance, [2]. Its main concept is that of the <u>ordinary partition relation</u>

$$a \to (b_i)^r_{i \in I} \tag{1}$$

which expresses the following statement: whenever $|A| = a$ and $f : [A]^r \to I$, there is $i \in I$ and $X \in [A]^{b_i}$ such that $f = i$ on $[X]^r$. If $|I| = n$ and $b_i = b$ for $i \in I$, then (1) is also written in the form

$$a \to (b)^r_n . \tag{2}$$

If α, β_i are ordinals then the relation $\alpha \to (\beta_i)^r_{i \in I}$ means that whenever $(A,<)$ is an ordered set of order type $tp(A,<) = \alpha$ and if $f : [A]^r \to I$, then there is $i \in I$ and $X \subseteq A$ such that $tp(X,<) = \beta_i$ and $f = i$ on $[X]^r$.

Theorem 1 asserts that

$$\aleph_0 \to (\aleph_0)^r_n \quad \text{for} \quad r,n < \aleph_0. \tag{3}$$

Ramsey also proved a finite version of (3):

Theorem 2.

Given $b,r,n < \aleph_0$, there is a cardinal $a < \aleph_0$ such that (2) holds.

Although several proofs of (3) are known, the following version of a proof might be of interest on account of its conciseness.
Proof of (3).

Let $1 \le r < \aleph_0$ and assume that

$$\aleph_0 \to (\aleph_0)^{r-1}_n \quad \text{for} \quad n < \aleph_0. \tag{4}$$

It suffices to deduce (3). Let $|A| = \aleph_0$; $|I| < \aleph_0$; $f : [A]^r \to I$. We define recursively x_ν, P_ν, Q_ν for $\nu < \omega$. Let $\nu < \omega$; $P_\nu \in [A]^\nu$; $Q_\nu \in [A \setminus P_\nu]^{\aleph_0}$. Then there is $Q'_\nu \in [Q_\nu]^{\aleph_0}$ such that $f(X \cup \{y\}) = f(X \cup \{z\})$ for $X \in [P_\nu]^{r-1}$ and $y,z \in Q'_\nu$. Choose $x_\nu \in Q'_\nu$ and put $P_{\nu+1} = P_\nu \cup \{x_\nu\}$; $Q_{\nu+1} = Q'_\nu \setminus \{x_\nu\}$. Apply this procedure repeatedly, for $\nu = 0,1,\ldots,\hat{\omega}$, starting with $P_0 = \emptyset$; $Q_0 = A$. In this way we obtain a set $B = \{x_0,x_1,\ldots,\hat{x}_\omega\}_{\neq}$ with the property that, for some function $g : [B]^{r-1} \to I$, we have $f(X \cup \{x_\alpha\}) = g(X)$ if $\alpha < \omega$ and $X \in [\{x_0,\ldots,\hat{x}_\alpha\}]^{r-1}$. By (4), there is a set $D \in [B]^{\aleph_0}$ such that g is constant on $[D]^{r-1}$. Then f is constant on $[D]^r$, and (3) is proved.

(b) I now state, without proof, some general results concerning ordinary partition relations. These, and more general propositions, are discussed in [2].

(i) Given $I \neq \emptyset$; $r < \aleph_0$ and arbitrary cardinals b_i for $i \in I$, there is a cardinal a such that (1) holds. If the <u>Generalized Continuum Hypothesis</u>

(GCH) $$2^c = c^+ \quad \text{for} \quad c \geq \aleph_0$$

is assumed then we have

$$\aleph_{\alpha+(r-1)} \to (\aleph_\alpha)^r_n$$

for $1 \leq r < \aleph_0$ and $n < \mathrm{cf}\,\aleph_\alpha$.

(ii) (A version of the positive stepping-up lemma)
If $1 \leq r < \aleph_0$; $a \geq \aleph_0$; $a \to (b_i)^r_{i\in I}$ and if GCH holds, then

$a^+ \to (b_i+1)^{r+1}_{i\in I}$.

The logical negation of a partition relation is obtained by writing $\nrightarrow$ instead of $\to$.

(iii) (A version of the negative stepping-up lemma)
Let $3 \leq r < \aleph_0$; $|I| \geq 2$; $b_i \geq \aleph_0$ for $i \in I$; $a \nrightarrow (b_i)^r_{i\in I}$. Then, assuming GCH, we have $a^+ \nrightarrow (b_i)^{r+1}_{i\in I}$.

The restriction in the ordinary relation to finite "exponents" r is natural, in view of the following result ([3],p.434).

Theorem 3.

If $a \geq 0$; $b \geq r \geq \aleph_0$; $n \geq 2$, then $a \nrightarrow (b)^r_n$.

(c) The <u>dual partition relation</u>

(5) $$a \to [b_i]^r_{i\in I} ,$$

also called the <u>square-bracket relation</u>, expresses the following condition. Whenever $|A| = a$ and $f : [A]^r \to I$, there is $i \in I$ and $X \in [A]^{b_i}$ such that $f \neq i$ on $[X]^r$. The next theorem establishes a link between the dual and the ordinary relation.

Theorem 4.

Let $|N| \geq 2$; $I_\nu \neq \emptyset$ for $\nu \in N$; $I_\mu \cap I_\nu = \emptyset$ for $\mu \neq \nu$; $I = I_N$; $b_i \leq c_\nu$ if $\nu \in N$ and $i \in I_\nu$; $a \to (c_\nu)^r_{\nu \in N}$. Then (5) holds.

By putting $N = I$; $I_\nu = \{\nu\}$ and $c_\nu = b_\nu$, we see that (1) implies (5), provided $|I| \geq 2$.

Proof of Theorem 4. Let $|A| = a$; $[A]^r = \bigcup(i \in I)K_i$; $K_i \cap K_j = \emptyset$ for $i \neq j$.

Put $L_\nu = K_{I \setminus I_\nu}$ for $\nu \in N$. Let $X \in [A]^r$. Then $X \in K_{i_0}$ for some $i_0 \in I$.

Next, $i_0 \in I_{\nu_0}$ for some $\nu_0 \in N$. Since $|N| \geq 2$, we can choose

$\nu_1 \in N \setminus \{\nu_0\}$. If $i_0 \in I_{\nu_1}$, then $i_0 \in I_{\nu_0} \cap I_{\nu_1} = \emptyset$ which is false. Hence

$i_0 \notin I_{\nu_1}$; $i_0 \in I \setminus I_{\nu_1}$; $X \in K_{i_0} \subseteq K_{I \setminus I_{\nu_1}} = L_{\nu_1} \subseteq L_N$; $[A]^r \subseteq L_N$. By

$a \to (c_\nu)^r_{\nu \in N}$ there is $\nu_2 \in N$ and $Y \in [A]^{c_{\nu_2}}$ such that $[Y]^r \subseteq L_{\nu_2}$. We

can choose $i \in I_{\nu_2}$. Then $|Y| = c_{\nu_2} \geq b_i$; $[Y]^r \cap K_i \subseteq L_{\nu_2} \cap K_i = K_{I \setminus I_{\nu_2}} \cap K_i$

$$\subseteq K_{I \setminus I_{\nu_2}} \cap K_{I_{\nu_2}} = \emptyset.$$

The existence of i and Y with the properties just deduced proves (5).

I state, without proof, a result for dual relations ([2], Theorem 22).

Theorem 5.

Assume GCH. Let $\mathrm{cf}\, a < a$ and $2 \leq r < \aleph_0$. Then

$$a \nrightarrow [a]^r_c \quad \text{for} \quad c \leq 2^{r-1}.$$

If, in addition, $\mathrm{cf}\, a = \aleph_0$, then

$$a \to [a]^r_c \quad \text{for} \quad c > 2^{r-1}.$$

(d) The <u>polarized</u> <u>partition</u> <u>relation</u>

$$\binom{a}{b} \to \binom{c_i}{d_i}_{i \in I}$$

expresses the following condition. Whenever $|A| = a$; $|B| = b$; $f : A \times B \to I$, then there is $i \in I$; $X \in [A]^{c_i}$; $Y \in [B]^{d_i}$ such that $f = i$ on $X \times Y$.

The ordinary relation with $r = 2$ has an obvious interpretation in terms of edge-colourings of a complete graph. Similarly, the polarized relation can be interpreted in terms of edge-colourings of a complete bipartite graph. I mention the following proposition ([2], Theorem 44).

Theorem 5.

Assuming GCH, we have, for $a, b \geq \aleph_0$, the relation $\binom{a}{b} \to \binom{a}{b}_2$ if, and only if,

$$\{a, a^+, \mathrm{cf}a, (\mathrm{cf}a)^+\} \cap \{b, b^+, \mathrm{cf}b, (\mathrm{cf}b)^+\} = \emptyset.$$

(e) Canonical partition relations. Let $(A,<)$ be an ordered set of order type $\mathrm{tp}(A,<) = \alpha$ and let $r < \aleph_0$. The symbol $\{x_0,\ldots,\hat{x}_r\}_<$ denotes the set $\{x_0,\ldots,\hat{x}_r\}$ and expresses the condition that $x_\alpha < x_\beta$ for $\alpha < \beta < r$.

A function $f : [A]^r \to I$ is called canonical if there are numbers $\varepsilon_0,\ldots,\hat{\varepsilon}_r \in \{0,1\}$ such that, for all choices of $\{x_0,\ldots,\hat{x}_r\}_<$, $\{y_0,\ldots,\hat{y}_r\}_< \subseteq A$, we have

$$f(\{x_0,\ldots,\hat{x}_r\}) = f(\{y_0,\ldots,\hat{y}_r\})$$

if, and only if, $x_\rho = y_\rho$ for every $\rho < r$ with $\varepsilon_\rho = 1$. We note that if $\alpha = \omega_\lambda$ and $(\varepsilon_0,\ldots,\hat{\varepsilon}_r) \neq (0,\ldots,\hat{0})$ then f takes $\aleph_\lambda$ different values. The canonical partition relation

$$\alpha \to (\beta)^r_{\mathrm{can}}$$

has the following meaning. Whenever $(A,<)$ is an ordered set of order type $\mathrm{tp}(A,<) = \alpha$, and $f : [A]^r \to I$ for some I, then there is $B \subseteq A$ with $\mathrm{tp}(B,<) = \beta$ such that the function f is canonical on $(B,<)$.

The main feature of canonical partition relations is the fact that, in contrast to all other kinds of partition relations, nothing is given about the number $|I|$ of "colours" in the given colouring $f : [A]^r \to I$. I mention two results and one conjecture, each linking ordinary and canonical relations.

Theorem 6 ([4], Theorem 46)

(i) Let $r < \aleph_0$ and $s = \binom{2r}{r}$. Let q be the number of distinct equivalence relations which can be defined on the set $\{0,1,\ldots,\hat{s}\}$. Let α, β be ordinals; $|\beta| > 2r$; $\alpha \to (\beta)^{2r}_q$. Then $\alpha \to (\beta)^r_{\mathrm{can}}$.

(ii) Assume GCH. Let α be an ordinal and $r < \aleph_0$. Then

$$\omega_{\alpha+2r+1} \rightarrow (\omega_\alpha + 2r+1)^{r+1}_{can} .$$

We recall that an infinite cardinal b is called (weakly) <u>inaccessible</u> if $b = cf\, b = b^-$.

Theorem 7.

Assume GCH. Suppose that the infinite cardinal $\aleph_\beta$ is not inaccessible. Then, for every ordinal α, we have

$$(6) \qquad \omega_\alpha \rightarrow (\omega_\beta)^2_{can}$$

if, and only if,

$$(7) \qquad \aleph_\alpha \rightarrow (\aleph_\beta)^2_n \quad \text{for every } n < \aleph_\beta.$$

It is easy to see that (6) implies (7). Let $|A| = \aleph_\alpha$ and $f : [A]^r \rightarrow I$; $|I| = n < \aleph_\beta$. Let $(A,<)$ be an order with $tp(A,<) = \omega_\alpha$. By (6) there is $X \subseteq A$ with $tp(X,<) = \omega_\beta$, such that f is canonical on $[X]^r$, say with the values $\varepsilon_0, \ldots, \hat{\varepsilon}_r \in \{0,1\}$ occurring in the definition of canonicity. Now, if $\varepsilon_\rho = 1$ for some ρ, then, by a remark made above, f would take $\aleph_\beta$ values, which contradicts the fact that $|I| < \aleph_\beta$. Hence all $\varepsilon_\rho = 0$, and so f is constant on $[X]^r$. This proves (7).

The proof of the implication (7) $\Rightarrow$ (6) is much more involved and makes use of some results of [2]. Since it is most likely that Theorem 7 remains true if the exponent 2 is replaced by any $r < \aleph_0$, we omit this proof. We state explicitly the conjecture which was just made:

<u>Conjecture</u>. Assume GCH. Let $r < \aleph_0$ and suppose that $\aleph_\beta$ is not inaccessible. Then $\omega_\alpha \rightarrow (\omega_\beta)^r_{can}$ holds if, and only if, $\aleph_\alpha \rightarrow (\aleph_\beta)^r_n$ for every $n < \aleph_\beta$.

PART II. Δ-SYSTEMS

The terms system and family are used as synonyms. A (a,b)-<u>family</u> is a family $(A_i : i \in I)$ which is such that $|I| = a$ and $|A_i| = b$ for $i \in I$. Expressions such as $(>a, \leq b)$-family are self-explanatory. The system $\mathfrak{A} = (A_i : i \in I)$ is called a <u>weak</u> (<u>strong</u>) Δ-system if for $i \neq j$ the cardinal $|A_i \cap A_j|$ (the set $A_i \cap A_j$) is indepentent of i,j. A Δ(c)-<u>system</u> is a Δ-system $(A_i : i \in I)$ with $|I| = c$.

Clearly, every strong Δ-system is weak. However, the system $(\{0,1\},\{0,2\},\{1,2\})$ is weak but not strong. The following simple proposition asserts that this particular system constitutes a critical case. Weak Δ-systems of more than three members, each of cardinal at most two, are automatically strong.

Theorem 8.

Every $(\geq 4, \leq 2)$-system which is a weak Δ-system is also a strong Δ-system.

Proof. Let the $(\geq 4, \leq 2)$-system $\mathfrak{A} = (A_i : i \in I)$ be weak Δ. We may assume that $0,1,2,3 \in I$ and that $|A_i \cap A_j| = p \leq 2$ for $i \neq j$.

Case 0. $p = 0$. Then $A_i \cap A_j = \emptyset$ for $i \neq j$.

Case 1. $p = 1$. If $|A_0| = 1$, then $|A_0 \cap A_i| = 1$ for $i \neq 0$; $A_0 \subseteq A_i$; $A_i \cap A_j = A_0$ for $i \neq j$. Now let $|A_i| = 2$ for all i. Let $A_0 = \{0,1\}$. Then, for $i \neq 0$, we have either $0 \in A_i$ or $1 \in A_i$. Without loss of generality we may assume that $0 \in A_1 \cap A_2$. We may further assume that $A_1 = \{0,2\}$ and $A_2 = \{0,3\}$.

Case 1a. $0 \notin A_{j_0}$ for some $j_0 \notin \{0,1,2\}$. Then $|A_i \cap A_{j_0}| = 1$ for $i \in \{0,1,2\}$ and therefore $1,2,3 \in A_{j_0}$, which contradicts $|A_{j_0}| = 2$.

Case 1b. $0 \in A_j$ for all $j \notin \{0,1,2\}$. Then $A_\alpha \cap A_\beta = \{0\}$ for $\alpha \neq \beta$, and $\mathfrak{A}$ is a strong Δ-system.

Case 2. $p = 2$. Then $A_\alpha \cap A_\beta = A_0$ for $\alpha \neq \beta$, and the theorem is proved.

Here is one further proposition which asserts that under certain conditions a weak Δ-system is automatically strong.

Theorem 9.

Let $\mathfrak{A} = (A_i : i \in I)$ be a weak Δ-system; $|A_i \cap A_j| = p < \aleph_0$ for $i \neq j$; $|A_i| = a_i$ for $i \in I$. Then $\mathfrak{A}$ is a strong Δ-system if (i) or (ii) holds where

(i) $a_i \leq n < \aleph_0$ for $i \in I$, and $|I| > 1 + n\binom{n}{p}$,

(ii) $|I| > 3 + \binom{a_i}{p}\binom{a_j}{p}\binom{a_k}{p}$ whenever $\{i,j,k\}_{\neq} \subseteq I$, where $\binom{a}{p} = |[A]^p|$ for $|A| = a \geq \aleph_0$.

Proof. The assertion relating to (i) is [5], Theorem 2. Now let (ii) hold. We may assume that, for some ordinal n, $I = \{0,1,\ldots,\hat{n}\}$. Put $D_{\alpha\beta} = A_\alpha \cap A_\beta$ for $\alpha, \beta < n$.

Case 1. Whenever $\alpha < \beta < \gamma < n$, then $(A_\alpha, A_\beta, A_\gamma)$ is strong Δ. Then

$D_{\alpha\beta} = D_{\alpha\gamma}$ whenever $\alpha \neq \beta$ and $\alpha \neq \gamma$. Now let $\alpha \neq \beta$ and $\gamma \neq \delta$. Then $D_{\alpha\beta} = D_{\gamma\delta}$, and $\mathfrak{A}$ is strong. For if $\alpha = \gamma$ then $D_{\alpha\beta} = D_{\gamma\beta} = D_{\gamma\delta}$, and if $\alpha \neq \gamma$ then $D_{\alpha\beta} = D_{\alpha\gamma} = D_{\delta\gamma}$.

Case 2. $n \geq 3$ and (A_0, A_1, A_2) is not strong Δ. Let $t < 3 \leq i < n$. Then $|D_{ti}| = p$. Hence
$|\{(D_{0i}, D_{1i}, D_{2i}) : 3 \leq i < n\}| \leq \binom{a_0}{p}\binom{a_1}{p}\binom{a_2}{p} = q$, say.

Case 2a. $|I \setminus \{0,1,2\}| > q$. Then there are ordinals i,j with $3 \leq i < j < n$ such that $D_{ti} = D_{tj}$ for all $t < 3$. Then, for $t < 3$,
$A_i \cap A_j \supseteq D_{ti} \cap D_{tj} = D_{ti}; |A_i \cap A_j| = p = |D_{ti}| < \aleph_0$.
Hence $A_i \cap A_j = D_{ti}$ for $t < 3$. If $s < t < 3$ then
$A_s \cap A_t \supseteq D_{si} \cap D_{ti} = A_i \cap A_j; |A_s \cap A_t| = p = |A_i \cap A_j| < \aleph_0$.
Hence $A_s \cap A_t = A_i \cap A_j$ for $s < t < 3$, which contradicts the fact that (A_0, A_1, A_2) is not strong Δ.

Case 2b. $|I \setminus \{0,1,2\}| \leq q$. Then $|I| \leq 3+q$, which contradicts the hypothesis (ii). The theorem is proved.

I shall now give a brief survey of some of the results that have been obtained. Let the relation $a \to \mathrm{st}\Delta(b,c)$ mean that every $(a,<b)$-system contains a strong $\Delta(c)$-subsystem. Given b,c, there is a least cardinal a, denoted by $f_{\mathrm{st}\Delta}(b,c)$, such that $a \to \mathrm{st}\Delta(b,c)$. Denote by $s(b,c)$ the supremum of all cardinals of the form $\sum_{\nu}(|\nu| < b)\prod_{\mu}(\mu < \nu)c_\mu$ (μ,ν range over ordinals), where the c_μ are any cardinals with $c_\mu < c$ for all μ.

Theorem 10.

Let $b \geq 2$; $c \geq 3$; $b+c \geq \aleph_0$. Then

$$f_{\mathrm{st}\Delta}(b,c) \in \{s(b,c), (s(b,c))^+\}.$$

This is part of [6] Theorem IV. In [6] one can find a complete discussion of the values taken by the function $f_{\mathrm{st}\Delta}$, both with and without the assumption of GCH. I should like to take this opportunity to remark that, as was kindly pointed out to the author by R.O.Davies, some errors have occurred in [6] in the computation of $f_{\mathrm{st}\Delta}$ under the assumption of GCH. These can, however, be easily corrected.

Theorem 11.

Let $m \geq \aleph_0$ and $n < m$. Let $\mathfrak{A} = (A_i : i \in I)$ be a (m^+, m)-system with $|A_i \cap A_j| < n$ for $i \neq j$. Then

(i) if $m^n = m$ then $\mathfrak{A}$ has a strong $\Delta(m^+)$-subsystem,

(ii) if $m^n > m$ and GCH holds , then $\mathfrak{A}$ has a strong $\Delta(p)$-subsystem for every $p < m$,

(iii) the statement (ii) becomes false if in (ii) we replace $p < m$ by $p \leq m$.

This is [5], Theorem 1.

Theorem 12.

If $a \geqslant 2$; $b \geqslant 1$; $a+b \geqslant \aleph_0$, then every $(>a^b, \leq b)$-system has a strong $\Delta(>a)$- subsystem.

This is part of [7], Theorem I.

Let the relation $(a,b) \rightarrow \mathrm{wk}\Delta(c)$ mean that every (a,b)-system has a weak $\Delta(c)$-subsystem. Given a,b, there is a least cardinal c, denoted by $\varphi_{\mathrm{wk}\Delta}(a,b)$, satisfying

$$(a,b) \nrightarrow \mathrm{wk}\Delta(c).$$

Theorem 13.

Assume GCH. If $a,b \geqslant \aleph_0$ then

$$\varphi_{\mathrm{wk}\Delta}(a,b) \in \{3, a^-, a, a^+\}.$$

This is proved in §7 of [5] where one can find a complete discussion of the function $\varphi_{\mathrm{wk}\Delta}$ on the assumption of GCH.

Clearly, it is desirable (i) to study, in analogy to the definition of the function $f_{\mathrm{st}\Delta}$, the function $\psi_{\mathrm{wk}\Delta}(a,b)$ whose value is the least cardinal c such that there is a $(a,<b)$-system without a weak $\Delta(c)$-subsystem, (ii) to carry out a discussion as in [5] but without assuming GCH.

Of various other problems I will only mention the following, which is very concrete and has claimed the attention of a fair number of authors.

Problem. Is there a positive integer c such that, for $1 \leq n < \aleph_0$, every (c^n, n)-system contains a strong $\Delta(3)$-subsystem?

REFERENCES

[1] F.P.Ramsey, On a problem of formal logic, Proc. London Math. Soc. (2) vol. 30 (1930), pp. 246-286.

[2] P.Erdös, A.Hajnal and R.Rado, Partition relations for cardinal numbers, Acta Math. Hungarian Acad. Scient. vol. XVI (1965), pp. 93-196.

[3] P.Erdös and R.Rado, Combinatorial theorems on classifications of subsets of a given set, Proc. London Math. Soc. (3) vol. 2 (1952), pp. 417-439.

[4] ———, A partition calculus in set theory, Bull. American Math. Soc. vol. 62 (1956), pp. 427-489.

[5] ———, Intersection theorems for systems of sets, Journ. London Math. Soc. vol. 35 (1960), pp. 85-90.

[6] ———, Intersection theorems for systems of sets (II), Journ. London Math. Soc. vol. 44 (1969), pp. 467-479.

[7] P.Erdös, E.C.Milner and R.Rado, Intersection theorems for systems of sets (III), Journ. Australian Math. Soc. vol. XVIII (1974), pp. 22-40.

ASPECTS OF FINITE LATTICES

R. Wille
Fachbereich Mathematik, Technische Hochschule,
Darmstadt, BRD

0. INTRODUCTION

There has been increasing interest in finite lattices in the last few years. This fact is witnessed by the large number of papers in current journals and by the fact that various reports on specific topics in this area have already appeared, namely CRAPO & ROTA [8], RIVAL [35], JÓNSSON & NATION [23], and, of course, certain sections in BIRKHOFF [7] and CRAWLEY & DILWORTH [10]. The following report can be understood in this light; it describes topics in the theory of finite lattices from an algebraic point of view concerning research work done mainly by members of the Technische Hochschule Darmstadt.

In this report we concentrate on two aspects of finite lattices: arithmetic and freeness. The first section discusses generalizations of Birkhoff's arithmetical theorem about the unique join representation of elements in finite distributive lattices. Especially, we consider the reconstruction of a finite lattice by its scaffolding and applications of this construction method to the arithmetic of polynomial functions on finite lattices. The second section deals with the description of finite lattices as free closures of partial lattices. For specific classes of partial lattices (e.g. partially ordered sets) there are characterizations of those partial lattices which freely generate finite lattices (with respect to the modular law or other lattice equations).

We would especially like to thank I. Rival for many interesting and fruitful discussions during the development of this report.

M. Aigner (ed.), Higher Combinatorics, 79-100. *All Rights Reserved.*

1. ARITHMETIC

In the theory of finite lattices there is an analogue of the fundamental theorem of arithmetic (the factorization of positive integers into primes) which is given by the following simple and elementary result.

THEOREM 0: Every element in a finite lattice has an irredundant representation as a join (meet) of join-irreducible (meet-irreducible) elements.

Let us recall that an element x in a finite lattice L is called join-irreducible (meet-irreducible) if $x \neq 0$ $(x \neq 1)$ and there are no elements y and z unequal x in L with $x = y \vee z$ $(x = y \wedge z)$. The partially ordered set of all join-irreducible (meet-irreducible) elements of L is denoted by J(L) (M(L)). A set X of elements forms an irredundant representation of its join (meet) if the join (meet) of each proper subset of X is unequal the join (meet) of X.

Although Theorem 0 can be easily verified, it contains the fundamental idea for the subdirect decomposition of algebras into subdirectly irreducible factors (s. BIRKHOFF [5]) which can be illustrated by the fact that an algebra with finite congruence lattice is subdirectly irreducible if and only if its identity is a meet-irreducible congruence relation. Conversely, the decomposition theorems in number theory and algebra have stimulated the search for arithmetical theorems in lattice theory (c.f. KLEIN [25]). The first final result which provides a satisfactory analysis of the arithmetic of finite distributive lattices is given by the following theorem.

THEOREM 1: (BIRKHOFF [4]) Every element in a finite distributive lattice L has a unique irredundant representation as a join (meet) of join-irreducible (meet-irreducible) elements; furthermore, L is isomorphic to the lattice of all order ideals (order filters) of J(L) (M(L)).

We recall that a subset X of a partially ordered set P is an order ideal (order filter) if $y \in X$ whenever $y \leq x$ $(y \geq x)$ for some $x \in X$. Obviously, the order ideals (order filters) of P form a complete lattice I(P) (F(P)).

In DILWORTH [13] it is shown that finite lattices satisfying the statement of Theorem 1 are not far from distributive lattices. Therefore further generalizations of Theorem 1 have to give up the uniqueness or the restriction to join-irreducible (meet-irreducible) elements. A celebrated generalization weakening the uniqueness is contained in the following result.

THEOREM 2: (KUROSCH [26], ORE [32]) Let L be a modular lattice.

If $a = x_1 \vee x_2 \vee \ldots \vee x_m$ and $a = y_1 \vee y_2 \vee \ldots \vee y_n$ are two irredundant join representations of an element a in L by join-irreducible elements, then for each x_i there exists a y_j such that $a = x_1 \vee x_2 \vee \ldots \vee x_{i-1} \vee y_j \vee x_{i+1} \vee \ldots \vee x_m$ is also an irredundant join representation of a ; especially, one obtains $m = n$.

The extent to which a converse of Theorem 2 holds is discussed in CRAWLEY [9]; again, a finite lattice satisfying the statement of Theorem 2 is in the neighbourhood of modular lattices.

For a generalization of the second part of Theorem 1 we need the definition of the normal completion of a partially ordered set P. Let $X^* := \{y \in P | y \geq x \text{ for all } x \in X\}$ and let $X_* := \{y \in P | y \leq x$ for all $x \in X\}$ for any subset X of P. Then the normal completion N(P) of P is the complete lattice of all sets $(X^*)_*$ with $X \subseteq P$ (c.f. CRAWLEY & DILWORTH [10], p.71).

THEOREM 3: (BANASCHEWSKI [1], SCHMIDT [39]) A finite lattice L is isomorphic to the normal completion of the partially ordered set $J(L) \cup M(L)$.

Reformulations and applications of Theorem 3 can be found in MARKOWSKY [28,29], URQUHART [49], and KELLY [24].

Although Theorem 2 and Theorem 3 are important generalizations, both theorems have lost essential parts of Theorem 1. Theorem 2 does not show how to reconstruct the lattices by the substructure of the specific elements under consideration wherefore the powerful duality between finite distributive lattices and finite partially ordered sets (c.f. PRIESTLEY [33,34]) has no counterpart. Theorem 3 provides such a duality, however the arithmetical content of Theorem 1 does not become apparent. In the following we consider a further generalization of Theorem 1 which may be understood as an attempt to get unique join representations by specific elements as well as a reconstruction by the substructure of those elements.

As it is mentioned above, we first have to generalize the concept of a join-irreducible element in finite lattices. The starting point for this generalization is the fact that an element of a finite lattice L is join-irreducible if and only if it covers exactly one element. We want to generalize this to elements which can cover more than one element however these covered elements are "densely" tied together.

An element x of a finite lattice L is called subirreducible (s. [56]) if $x \neq 0$ and there is a pair of elements $a > b$ in L such that for each element y covered by x there exist $c_1, c_2, \ldots, c_{2n} \in L$ with

$$(\ldots((((x \wedge c_1) \vee c_2) \wedge c_3) \vee c_4) \wedge \ldots c_{2n-1}) \vee c_{2n} = a \quad \text{and}$$

$$(\ldots((((y \wedge c_1) \vee c_2) \wedge c_3) \vee c_4) \wedge \ldots c_{2n-1}) \vee c_{2n} = b$$

(c.f. SCHMIDT [38],p.69). Obviously, any join-irreducible element x of L is subirreducible because, if y is the unique element covered by x, we may choose $a := x$, $b := y$, $c_1 := x$, and $c_2 := y$. for many purposes the following external characterization of subirreducible elements of L is very helpful: An element x of L is subirreducible if and only if there exists a surjective homomorphism φ from L onto a subdirectly irreducible lattice with $x = \bigwedge \varphi^{-1}\varphi x$ (s. [56]). The set of all subirreducible elements of L together wit the restriction of the join operation to this set forms a partial join-semilattice G(L) which is called the <u>scaffolding</u> of L. We recall that a <u>partial (join-) semilattice</u> $(\underset{\sim}{S}, \underset{\sim}{\vee})$ is defined by a (join-) semilattice $(S, \vee)$ where $\underset{\sim}{S} \subseteq S$ and $x \underset{\sim}{\vee} y = z$ for $x,y,z \in \underset{\sim}{S}$ if and only if $x \vee y = z$ (one often writes $\underset{\sim}{S}$ for $(\underset{\sim}{S}, \underset{\sim}{\vee})$ and $\vee$ for $\underset{\sim}{\vee}$) (c.f. GRÄTZER [18],p.48). Obviously, the restriction of the partial order of $(S, \vee)$ can be defined on $\underset{\sim}{S}$ by the partial operation $\underset{\sim}{\vee}$. Therefore we may internally define an <u>ideal</u> X of $(\underset{\sim}{S}, \underset{\sim}{\vee})$ as a subset of $\underset{\sim}{S}$ such that $z \in X$ whenever $z \leqslant x$ or $z = x \underset{\sim}{\vee} y$ for some $x,y \in X$. The lattice of all ideals of $(\underset{\sim}{S}, \underset{\sim}{\vee})$ is denoted by $I(\underset{\sim}{S})$. The set of maximal elements of an ideal in $(\underset{\sim}{S}, \underset{\sim}{\vee})$ is called a <u>closed subset</u> of $\underset{\sim}{S}$. It can be easily seen that the scaffolding of a finite distributive lattice L is exactly the partially ordered set J(L) and that the closed subsets of J(L) are the antichains of J(L) (s. [56]).

Before stating the **general** representation theorem we wish to illustrate the introduced concepts by the modular lattice FM(3) freely generated by three elements x,y and z (FM(3) was first described by DEDEKIND [12]).

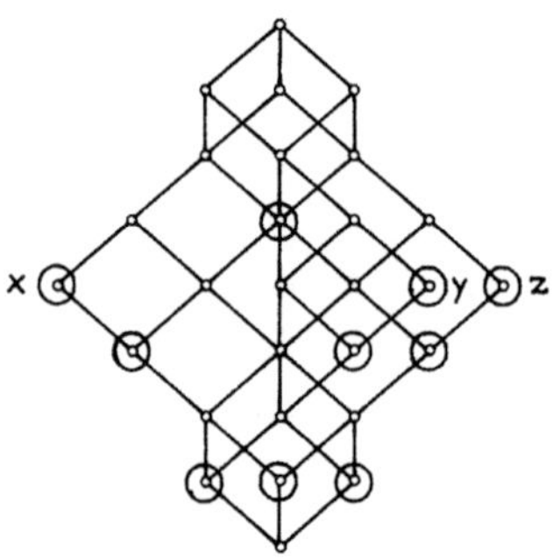

FM(3)

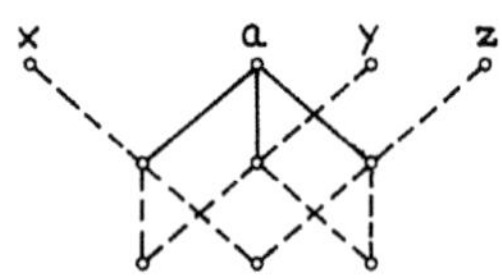

G(FM(3))

In the left diagram of FM(3) the subirreducible elements are indicated by big circles. $(x \vee y) \wedge (x \vee z) \wedge (y \vee z)$ is the only subirreducible element which is not join-irreducible; for checking its subirreducibility we may choose $a := (x \vee y) \wedge (x \vee z) \wedge (y \vee z)$, $b := (x \wedge z) \vee (y \wedge a)$, $c_2 := b$, and either $c_1 := (x \wedge y) \vee (z \wedge a)$ for the covered element $(y \wedge z) \vee (x \wedge a)$ or $c_1 := a$ for b or $c_1 := (y \wedge z) \vee (x \wedge a)$ for $(x \wedge y) \vee (z \wedge a)$. In the right diagram the partial order of G(FM(3)) is described by all lines where the non-trivial joins are indicated only by the straight lines. An example of a closed subset of G(FM(3)) is given by {x,z,a}; but {x,z} is not closed.

THEOREM 4: [56] Every element in a finite lattice L has a unique representation as the join of a closed set of subirreducible elements; furthermore, L is isomorphic to the lattice of all ideals of its scaffolding G(L).

As Theorem 1 for finite distributive lattices, Theorem 4 yields a reduction for the description of finite lattices and their arithmetic, especially for finite lattices which are large with respect to their subdirectly irreducible factors. This reduction may be visualized by the lattice $FN_5(3)$ freely generated by three elements with respect to the lattice equations valid in the 5-element non-modular lattice N_5 (s. WATERMANN [50]); the cardinalities of this lattice and its scaffolding are $|FN_5(3)| = 99$ and $|G(FN_5(3))| = 15$ (s.[56]).

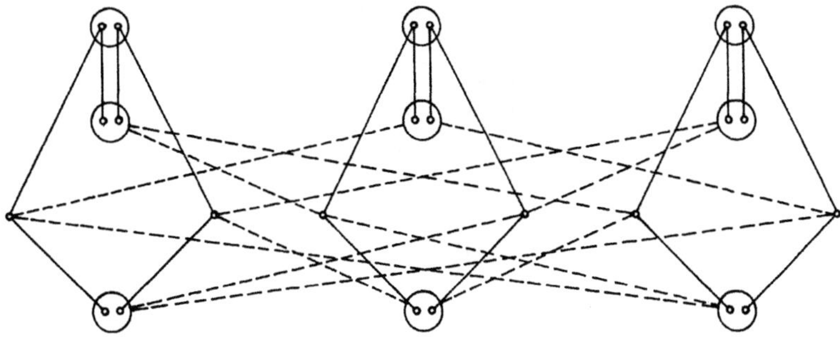

$G(FN_5(3))$

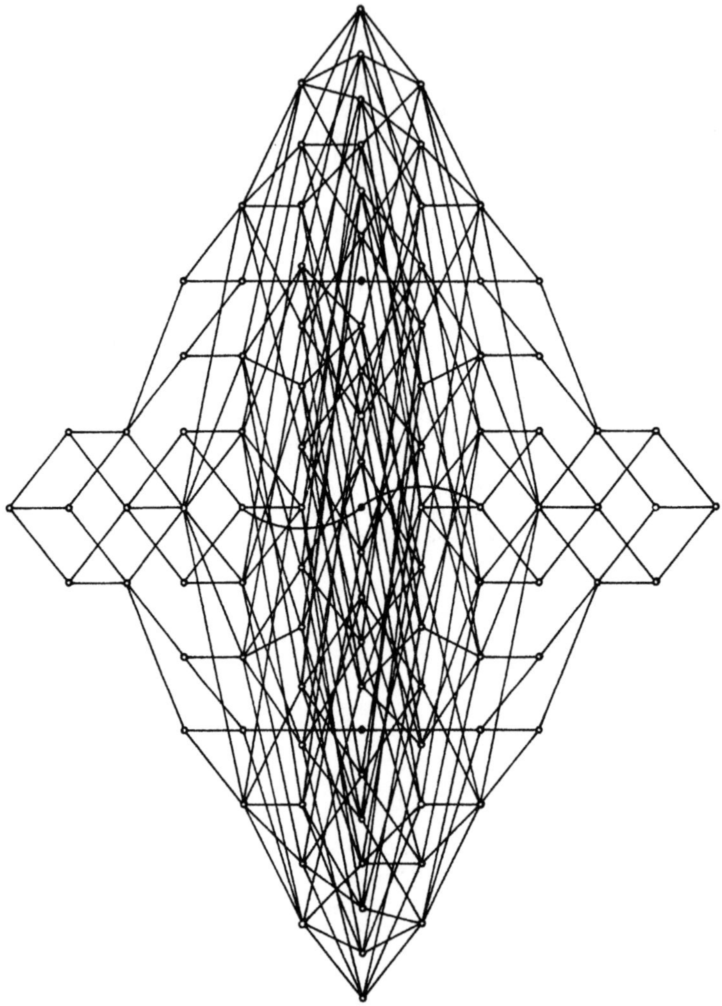

$FN_5(3)$

As another example we consider the modular lattice $FM(\mathbf{1}+\mathbf{1}+\mathbf{n})$ freely generated by two elements e_1, e_2 and an n-element chain $e_3 < e_4 < \ldots < e_{n+2}$. For $n = 5$ we get $|FM(\mathbf{1}+\mathbf{1}+\mathbf{5})| = 12134$ and $|G(FM(\mathbf{1}+\mathbf{1}+\mathbf{5}))| = 82$ (s. [53,56]) wherefore only the scaffolding of $FM(\mathbf{1}+\mathbf{1}+\mathbf{n})$ is shown.

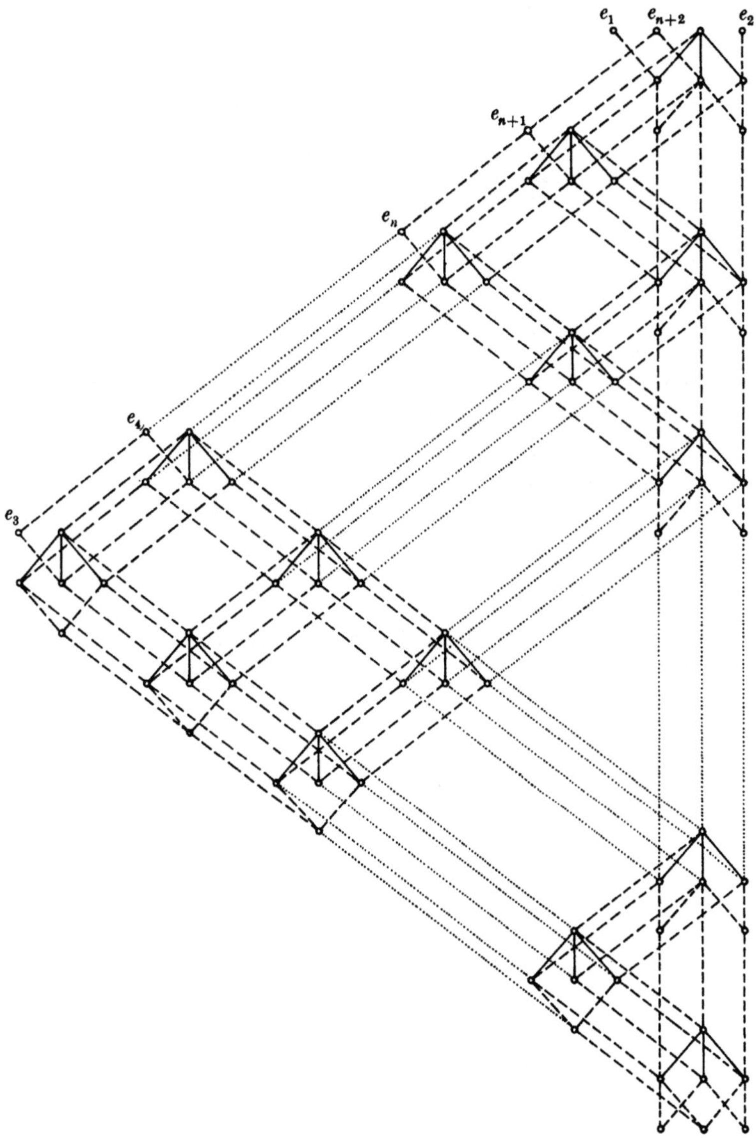

$G(FM[1 + 1 + n])$

How arithmetical questions can be handled via the scaffolding is shown by an examination of the complexity of lattice terms which one needs to describe the elements of $FM(\mathbf{1}+\mathbf{1}+\mathbf{n})$ by the generators. Obviously, the join-irreducible elements of a finite lattice are exactly the join-irreducible elements of its scaffolding. Using this fact, we can easily check with the above figure that the join-irreducible elements of $FM(\mathbf{1}+\mathbf{1}+\mathbf{n})$ can be represented in the form $e_i \wedge ((e_j \wedge e_k) \vee (e_j \wedge e_l))$ $(1 \leq i,j,k,l \leq n+2)$ which improves a claim in SCHÜTZENBERGER [41].

In working with Theorem 4 it becomes desirable to make the duality between finite lattices and their scaffoldings more useful. In a first step this requires a characterization of scaffoldings as partial join-semilattices (c.f. SCHMIDT [40]). We consider in the following an approach of GANTER & POGUNTKE & WILLE [16] which was modified and generalized to arbitrary lattices by adding topological conditions in GIERZ & KEIMEL [17].

The fundamental observation is that for a homomorphism φ from a finite lattice onto a subdirectly irreducible lattice S we get an embedding $\underline{\varphi}$ from the join-semilattice $S^{\vee} := (S\setminus\{0\},\vee)$ into the scaffolding G(L) by $\underline{\varphi}x := \bigwedge \varphi^{-1}x$ $(x \in S\setminus\{0\})$. Therefore G(L) is the union of the join-semilattices $\underline{\varphi}S^{\vee}$, called the <u>components</u> of G(L), where $S := L/\Theta$ is a subdirectly irreducible factor and φ is the canonical homomorphism from L onto L/Θ. This motivates the following definition: A <u>partial $\mathfrak{K}$-semilattice with components</u> $(L_i,\vee_i)$ $(i \in I)$ is a partial join-semilattice $(P,\vee)$ satisfying the following conditions:

(1) $P = \bigcup_{i\in I} L_i$.

(2) $(L_i,\vee_i) \in \mathfrak{K}$ for all $i \in I$ where $\mathfrak{K}$ is a class of join-semilattices closed under the formation of isomorphic images.

(3) $x \vee y = z$ for $x \neq z \neq y$ in $(P,\vee)$ if and only if there exist an $i \in I$ and $\underline{x},\underline{y} \in L_i$ with $\underline{x} \leq x \leq z \geq y \geq \underline{y}$, $z \in L_i$, and $\underline{x} \vee_i \underline{y} = z$.

<u>THEOREM 5</u>: (GANTER & POGUNTKE & WILLE [16]) Let $\mathfrak{T}$ be an isomorphically closed class of subdirectly irreducible finite lattices. A finite partial join-semilattice P is isomorphic to the scaffolding of a finite lattice L with $L/\Theta \in \mathfrak{T}$ for all $\Theta \in M(\mathcal{C}(L))$ (where $\mathcal{C}(L)$ is the congruence lattice of L) if and only if P is a partial $\{S^{\vee} | S \in \mathfrak{T}\}$-semilattice with components S_Θ $(\Theta \in M(\mathcal{C}(L)))$ such that there are isomorphisms $\xi_\Theta : S_\Theta \to (L/\Theta)^{\vee}$ yielding join-preserving maps $\alpha_\Theta : P \to L/\Theta$ by $\alpha_\Theta x := \bigvee\{\xi_\Theta y | y \in S_\Theta \text{ with } y \leq x\}$ for $x \in P$ $(\Theta \in M(\mathcal{C}(L)))$.

Theorem 5 shows that a partial $\mathfrak{K}$-semilattice is the scaffolding of a finite lattice if and only if the union of any two of its components is a scaffolding of a finite lattice. The next theorem demonstrates how this characterization can be used to describe all scaffoldings of finite lattices in a specific equationally defined class. We consider the class $\mathfrak{M}_3$ consisting of all lattices which satisfy all lattice equations valid in the 5-element non-distributive modular lattice M_3. $\mathfrak{M}_3$ is the smallest equationally defined class of modular lattices properly extending the class $\mathfrak{D}$ of all distributive lattices (c.f. JÓNSSON [22]); $\mathfrak{D}$ consists of all lattices satisfying all lattice equations valid in the 2-element lattice D_2 (s. BIRKHOFF [2], STONE [44]).

<u>THEOREM 6</u>: (GANTER & POGUNTKE & WILLE [16]) A finite partial $\mathfrak{K}$-semilattice with at least two components is the scaffolding of a finite lattice in $\mathfrak{M}_3$ if and only if the union of any two of its components is described by one of the following diagrams:

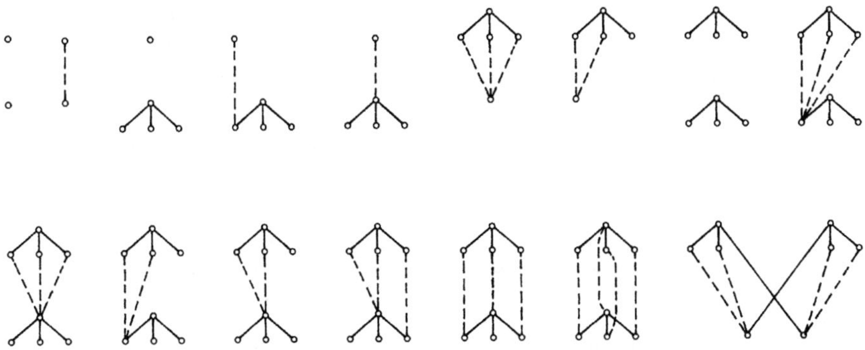

What really makes the concept of a scaffolding fruitful is the existence of construction methods for building up the scaffolding of a finite subdirect product by the scaffoldings of its factors. Such methods are developed for a more general situation in [56]. For finite lattices the initial question is how to compute effectively a finite subdirect product which is given by its factors and a set of generators. Because of the absorption laws calculating joins and meets can be very redundant. This difficulty can be avoided by a construction of specific elements for join-representations or by a construction of a partial join-semilattice whose ideal lattice is isomorphic to the subdirect product.

An <u>n-lattice</u> is a pair (L,α) where L is a lattice and α is a map from $\{1,2,\ldots,n\}$ onto a generating subset of L. An n-lattice (L,α) is called the <u>n-product</u> of the n-lattices (L_i,α_i) $(i\in I)$ if L is a sublattice of the direct product $\bigtimes_{i\in I} L_i$ and $\alpha k = (\alpha_i k \mid i\in I)$

for all $k\in\{1,2,\ldots,n\}$ (s. [52]). A finite n-product (L,α) can be constructed by its factors (L_i,α_i) $(i \in I)$ by the following two procedures (s. [56]).

Construction I:

1. Determine the greatest join-preserving maps $\alpha_{ij} : L_j \to L_i$ with $\alpha_{ij}\alpha_j k \leq \alpha_i k$ for all $k \in \{1,2,\ldots,n\}$ $(i,j \in I)$.

2. Form the subset $G(\alpha_{ij} | i,j \in I) := \{(\alpha_{ij}x | i \in I) | j \in I \text{ and } x \in L_j\}$ of $\bigtimes_{i\in I} L_i$.

3. Compute (L,α) by $L = \{\bigvee A | A \subseteq G(\alpha_{ij} | i,j \in I)\}$ and by $\alpha k = (\alpha_i k | i \in I)$ for all $k \in \{1,2,\ldots,n\}$.

Construction II:

1. Determine the greatest join-preserving maps $\alpha_{ij} : L_j \to L_i$ with $\alpha_{ij}\alpha_j k \leq \alpha_i k$ for all $k \in \{1,2,\ldots,n\}$ $(i,j \in I)$.

2. Form the partial join-semilattice $H(\alpha_{ij} | i,j \in I) := ((\bigcup_{i\in I}\{i\}\times L_i^{\vee})/\theta, \vee)$ where $(i,x)\theta(j,y) \Leftrightarrow x \leq \alpha_{ij}y, y \leq \alpha_{ji}x$ and $(\overline{i,x}) \vee (\overline{j,y}) = (\overline{k,z}) \Leftrightarrow x \leq \alpha_{ik}z, y \leq \alpha_{jk}z, z = \alpha_{ki}x \vee_k \alpha_{kj}y$ for $i,j,k \in I$ and $x \in L_i^{\vee}, y \in L_j^{\vee}, z \in L_k^{\vee}$.

3. Compute (L,α) by $L \cong I(H(\alpha_{ij} | i,j \in I))$ and by $\tilde{\alpha}k := \{(\overline{i,x}) | i \in I, x \in L_i^{\vee}$ with $x \leq \alpha_i k\}$ for all $k \in \{1,2,\ldots,n\}$.

Construction I and Construction II have been used to determine finite lattices freely generated by partial lattices with respect to certain lattice equations as $FN_5(3)$ or $FM(\mathbf{1}+\mathbf{1}+\mathbf{n})$ (s. [53,56]). As another type of applications, we discuss in the following how these constructions may help to analyse the arithmetic of polynomial functions on finite lattices. In showing that every lattice equation valid in N_5 can be derived by the lattice axioms and five other specific lattice equations valid in N_5, McKENZIE [30] has given an impressive demonstration of the difficulties which can arise in a pure syntactical study of lattice terms. Such difficulties may indicate that a more semantical approach is preferable for the general study of polynomial functions which are functions described by lattice terms with constant from an underlying lattice (c.f. LAUSCH & NÖBAUER [27]).

For a lattice L (unary) polynomial functions are the elements of the sublattice Q(L) which is generated by the identity function and the constant functions in the lattice of all functions from L into itself. The described constructions can be applied to determine Q(L) for a finite lattice L because there are n-lattices $(Q(L),\alpha)$ and (L,α_x) $(x \in L)$ with $n = |L| + 1$ such that $(Q(L),\alpha)$ is the n-product of the n-lattices (L,α_x) $(x \in L)$. This shall be examplified by the construction of $Q(N_5)$.

First we define $\alpha : \{1,2,\dots,6\} \to Q(N_5)$ in the way that $\alpha 1$ is the identity function and the $\alpha k (2 \leq k \leq 6)$ are the constant functions with $\alpha 2 < \alpha 4 < \alpha 6$ and $\alpha 2 < \alpha 3 < \alpha 5 < \alpha 6$. To get $(Q(N_5),\alpha)$ as 6-product of the 6-lattices (N_5,α_x), the $\alpha_x : \{1,2,\dots,6\} \to N_5$ $(x \in N_5)$ have to be defined by $\alpha_x 1 := x$ and $\alpha_x k := \alpha k x$ for $2 \leq k \leq 6$. For both constructions we must determine the greatest join-preserving maps $\alpha_{xy} : N_5 \to N_5$ with $\alpha_{xy} y \leq x$ and $\alpha_{xy} z \leq z$ for all $z \in N_5$ $(x,y \in N_5)$. Obviously, α_{xy} is the identity function if $y \leq x$. The maps α_{xy} with $y \nleq x$ are described by the following diagram which only indicates $z \mapsto \alpha_{xy} z$ for elements z with $\alpha_{xy} z \neq 0$ and $z = \bigwedge \alpha_{xy}^{-1} \alpha_{xy} z$.

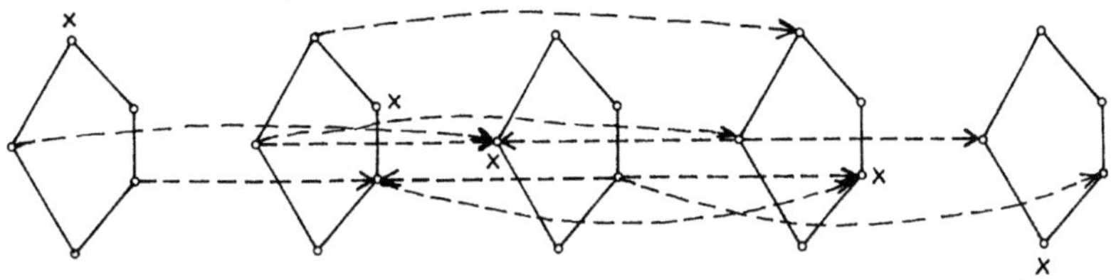

Continuing Construction I, we get as elements of $G(\alpha_{xy} | x,y \in N_5)$ $(= G(Q(N_5)))$

(33232), (33333), (42422), (44444), (53232), (53533), (55232),
(55252), (55555), (63432), (63633), (66464) and (66666).

where $(x_6 x_5 x_4 x_3 x_2)$ describes the function f with $f\alpha k = \alpha x_k$ for $2 \leq k \leq 6$. Now, taking the join of all subsets of $G(\alpha_{xy} | x,y \in N_5)$, the construction ends with the 24 elements of $Q(N_5)$ (c.f. DORNINGER & WIESENBAUER [15], Satz 6). For construction II we have to form the partial join-semilattice $H(\alpha_{xy} | x,y \in N_5)$ which is described by the diagram on the left. Finally, the ideal lattice of $H(\alpha_{xy} | x,y \in N_5)$ isomorphic to $Q(N_5)$ is shown on the right.

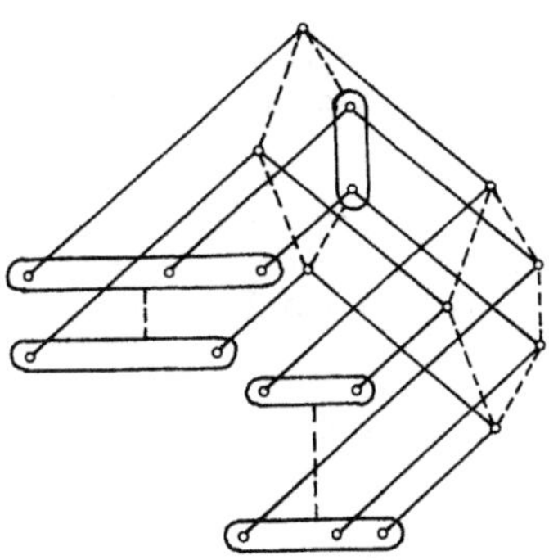

$G(Q(N_5))$

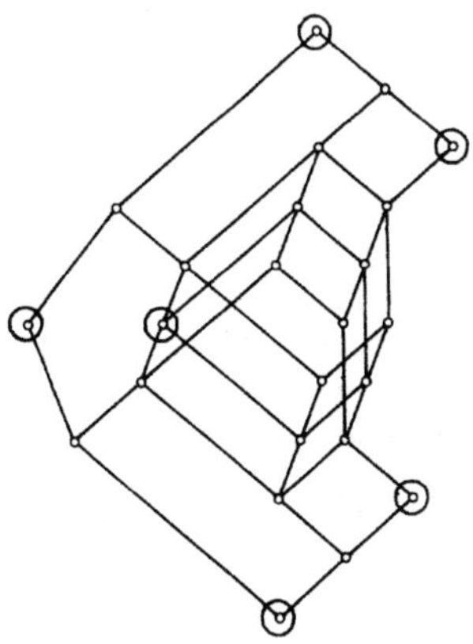

$Q(N_5)$

The construction of $Q(N_5)$ indicates that there is a general method for clarifying the arithmetic of polynomial functions on a finite lattice L. Let us define α_{xy} to be the greatest join-preserving map from L into itself with $\alpha_{xy}y \leqslant x$ and $\alpha_{xy}z \leqslant z$ for all $z \in L$ ($x,y \in L$). For $y,z \in L$ we define $f_y^z : L \longrightarrow L$ by $f_y^z x := \alpha_{xy}z$ for all $x \in L$.

THEOREM 7: [60,61] If L is a finite lattice and $y,z \in L$, then f_y^z is a polynomial function on L; furthermore, $f = \bigvee\{f_y^z \mid y \in L \text{ and } fy = z\}$ for each $f \in Q(L)$.

Theorem 7 can be used for the characterization of finite lattices which are rich in polynomial functions. A lattice L is called order polynomial complete if every order-preserving function from L into itself is a polynomial function (s. SCHWEIGERT [42]).

THEOREM 8: [60] A finite lattice L is order polynomial complete if and only if the identity function and the constant function to 0 are the only join-preserving maps $\delta : L \longrightarrow L$ with $\delta x \leqslant x$ for all $x \in L$.

From Theorem 8 and Construction II we derive the following characterization of the scaffolding of Q(L) for a finite order polynomial complete lattice L.

THEOREM 9: Let L be a finite order polynomial complete lattice. Then the scaffolding of Q(L) is isomorphic to the partial join-semilattice $(L \times L^{\vee}, \underset{\sim}{\vee})$ where $(x_1,x_2) \underset{\sim}{\vee} (y_1,y_2) = (z_1,z_2)$ if and only if $x_1 = y_1 = z_1$, $x_2 \vee y_2 = z_2$ or $x_1 \geqslant y_1 = z_1$, $x_2 \leqslant y_2 = z_2$ or $y_1 \geqslant x_1 = z_1$, $y_2 \leqslant x_2 = z_2$.

That Theorem 9 clarifies for instance the arithmetic of polynomial functions on irreducible finite geometric lattices becomes clear by the following theorem which is also a consequence of Theorem 8.

THEOREM 10: [60] A finite lattice L whose greatest element is the join of atoms is order polynomial complete if and only if L is simple.

The section will be closed by a result showing that the subspace lattices of (irreducible) finite projective geometries are (up to isomorphism) those finite modular lattices which are rich in polynomial functions.

THEOREM 11: [60] A finite modular lattice L is order polynomial complete if and only if L is irreducible and geometric.

2. FINITENESS AND FREENESS

Groups and other algebraic structures are often described by sets of relations with respect to certain generators. In this section we want to discuss that descripton method for finite lattices.

For example, it can be easily checked that the relations $(a \vee b) \wedge c = b$ and $b \vee c = c$ for the generators a,b, and c yield a lattice isomorphic to the direct product of the 2-element and the 3-element chain.

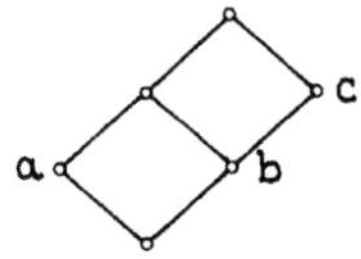

This lattice can also be understood as the lattice freely generated by the partial lattice which is formed by the elements a,b,c, and $a \vee b$. The change from relations to partial lattices makes the defining conditions more pictorial and is therefore often used in lattice theory. In general, a set of relations with respect to generators $x_1, x_2, \ldots, x_n$ determines a partial lattice as follows: we take the set P of elements described by all subterms of the relations in the lattice FL(n) freely generated by the $x_1, \ldots, x_n$ and make P to a partial lattice in adding the restrictions of the join and meet operation to P (c.f. GRÄTZER [18], p.48). This partial lattice freely generates the lattice described by the given relations. We recall that the lattice FL(P) freely generated by the partial lattice P is (up to isomorphism) characterized by the following condition: FL(P) is generated by P and for each homomorphism φ from P into a lattice L there exists a homomorphism $\varphi : FL(P) \longrightarrow L$ extending φ (c.f. GRÄTZER [18], Section 5).

Let us start the discussion of the description method by free generation with the tautological statement that every finite lattice is freely generated by a finite partial lattice. Of course, any finite lattice is at least freely generated by itself. But the really interesting situation occurs when there is a larger difference between the lattice and its generating partial lattice such that we get an effective reduction in the description of the lattice as free closure of a partial lattice. The question when a finite lattice can effectively be described by a partial lattice could be answered if one is able to solve the following problem.

PROBLEM: For which finite partial lattices P is FL(P) finite?

The less structure a finite partial lattice P possesses the larger becomes the difference between the cardinalities of FL(P) and P. Thus, it is natural to start with the totally unordered sets, i.e. to consider the free lattices FL(n). Unfortunately, we do not get much for our problem because $|FL(1)| = 1$, $|FL(2)| = 4$, and $|FL(n)| = \aleph_0$ for all $n \geqslant 3$ (s. BIRKHOFF [3]). But also the disjoint union $\mathfrak{n}_1 + \mathfrak{n}_2 + \ldots + \mathfrak{n}_m$ of chains $C_1, C_2, \ldots, C_m$ with $|C_i| = n_i \geqslant 1$ $(1 \leq i \leq m)$ do not yield much more because $|FL(\mathfrak{n}_1 + \mathfrak{n}_2 + \ldots + \mathfrak{n}_m)| = \aleph_0$

if $m \geq 3$ or $n_1 \cdot n_2 \geq 4$ (s. SORKIN [43]). Diagrams of the infinite lattices FL($\mathbb{1}+\mathbb{4}$) and FL($\mathbb{2}+\mathbb{2}$) can be found in ROLF [36]. The 9-element lattice FL($\mathbb{1}+\mathbb{2}$) and the 20-element lattice FL($\mathbb{1}+\mathbb{3}$) have the following diagrams (s. SORKIN [43]):

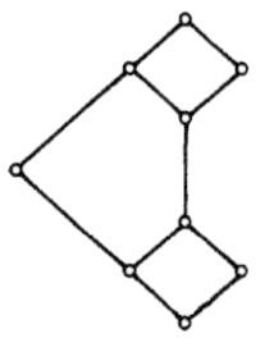

FL($\mathbb{1}+\mathbb{2}$)

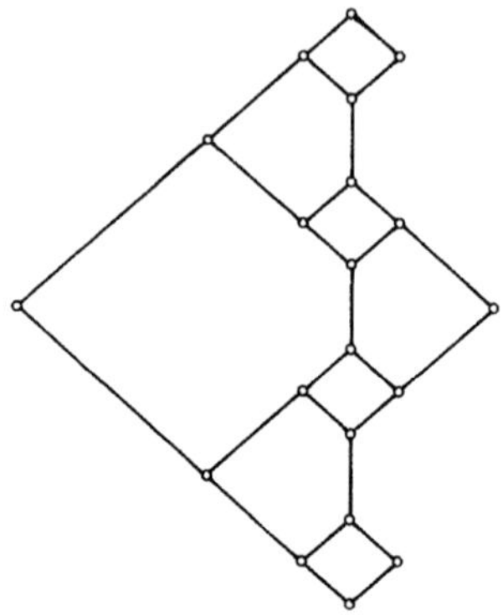

FL($\mathbb{1}+\mathbb{3}$)

In a further step we discuss finite partially ordered sets which are considered as partial lattices in the way that $x \vee y = z$ if and only if $x \lessdot y = z$ or $y \lessdot x = z$ and for $\wedge$ dually. In generalizing the mentioned results of Birkhoff and Sorkin, the following theorem solves the problem for finite partially ordered sets.

THEOREM 12: [59] For a finite partially ordered set P the following conditions are equivalent:
(i) FL(P) is finite.
(ii) FL(P) is a subdirect product of lattices described by the following diagrams:

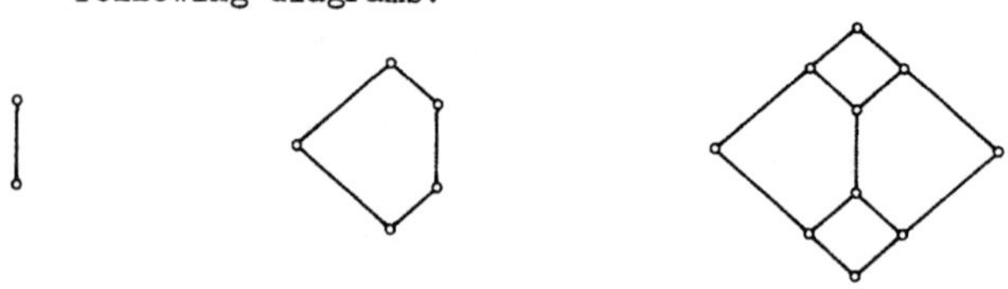

(iii) P does not contain an isomorphic copy of $\mathbb{1}+\mathbb{1}+\mathbb{1}$, $\mathbb{2}+\mathbb{2}$, and $\mathbb{1}+\mathbb{4}$.

Condition (ii) makes it possible to determine the finite lattices FL(P) by the construction methods of our first section. We first have to find a maximal family of homorphisms $\alpha_i : P \to L_i$ ($i \in I$) where L_i is a subdirectly irreducible factor of FL(P) (described in (ii)) such that $\alpha_i P$ generates L_i and there is no isomorphism $\psi : L_i \longrightarrow L_j$ $(i \neq j)$ with $\psi\alpha_i = \alpha_j$. Then we construct the n-product (L,α) with the factors $(L_i,\tilde{\alpha}_i)$ $(i \in I)$ and $n = |P|$ where $\tilde{\alpha}_i = \alpha_i\nu$ for a fixed bijection $\nu : \{1,2,\ldots,n\} \longrightarrow P$. Now, FL(P) is determined because of $L \cong FL(P)$. In [59], the finite lattices FL(P) are completely described with help of a recursive construction of the finite partially ordered sets P satisfying condition (iii). Instead of recalling the full description of those partially ordered sets, which are ordinal sums of ordinally irreducible partially ordered sets $P(n_1,n_2,\ldots,n_m)$, some examples of finite partially ordered sets satisfying (iii) are given.

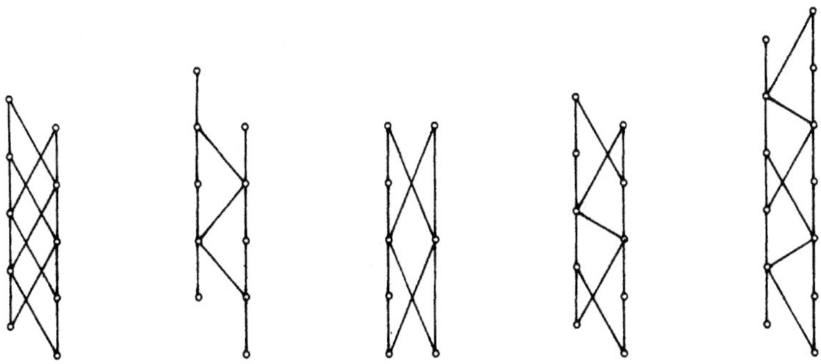

P(8) P(0,4) P(2,1,2) P(2,2,2) P(1,2,1,2,1)

At the moment, nothing more is known about solutions to our problem besides some small isolated results which are more or less folklore. To get just one example, we notice that the 8-element boolean lattice is freely generated by the partial lattice formed by its atoms and coatoms (c.f. GRÄTZER [18], Lemma 5.9).

For the study of finiteness and freeness in the given context it is fundamental that FL(P) is finite if and only if there are (up to isomorphism) only finitely many finite subdirectly irreducible lattices generated by a homomorphic image of P and there is no infinite subdirectly irreducible lattice generated by a homomorphic image of P. In this light the following modification of Theorem 12 may be of some help for further examinations of lattices generated by partial lattices.

<u>Theorem 13</u> [58] For a finite partially ordered set P the following conditions are equivalent:

(i) There are (up to isomorphism) only finitely many simple lattices generated by a homomorphic image of P.

(ii) The 1-element lattice D_1, the 2-element lattice D_2, and the 5-element lattice M_3 are (up to isomorphism) the only simple lattices generated by a homomorphic image of P.

(iii) P does not contain an isomorphic copy of $\mathbb{1}+\mathbb{1}+\mathbb{1}+\mathbb{1}$, $\mathbb{1}+\mathbb{1}+\mathbb{2}$,

and $\mathbb{1}+\mathbb{K}_2$ where $\mathbb{1}+\mathbb{K}_2$ is described by 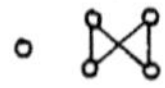

COROLLARY: A simple lattice generated by at most three elements is isomorphic to D_1, D_2 or M_3.

Besides other examples we obtain infinite lists of non-isomorphic finite simple lattice generated by isomorphic copies of the "critical" partially ordered sets $\mathbb{1}+\mathbb{1}+\mathbb{1}+\mathbb{1}$ and $\mathbb{1}+\mathbb{1}+\mathbb{2}$ by the following remarkable result.

THEOREM 14: (STRIETZ [45,46]) The lattice of all partitions on an n-element set has a subset of generators isomorphic to $\mathbb{1}+\mathbb{1}+\mathbb{1}+\mathbb{1}$ for all $n \geq 4$ and to $\mathbb{1}+\mathbb{1}+\mathbb{2}$ for all $n \geq 10$.

Free closures of partial lattices are also considered in other equationally defined classes of lattices than the class of all lattices. For an equationally defined class $\mathcal{L}$ of lattices the lattice $F(P,\mathcal{L})$ $\mathcal{L}$-freely generated by the partial lattice P is (up to isomorphism) characterized by the following condition: $F(P,\mathcal{L})$ is contained in $\mathcal{L}$, there is a homomorphism ι from P onto a generating subset of $F(P,\mathcal{L})$ and for each homomorphism φ from P into a lattice L of $\mathcal{L}$ there exists a homomphism $\hat{\varphi}: F(P,\mathcal{L}) \to L$ with $\hat{\varphi}\iota = \varphi$ (c.f. GRÄTZER [18], Section 5).

In the following we consider the question which finite partial lattices P have finite $\mathcal{L}$-free closures $F(P,\mathcal{L})$ mainly for the class $\mathfrak{M}$ of all modular lattices. Usually, $F(P,\mathfrak{M})$ is denoted by FM(P). For totally unordered sets, i.e. for the free modular lattices FM(n), we have $|FM(1)| = 1$, $|FM(2)| = 2$, $|FM(3)| = 28$, and $|FM(n)| = \aleph_0$ for all $n \geq 4$ by DEDEKIND [12] and BIRKHOFF [2]. In generalizing these results and further results of BIRKHOFF [6;p.72], TAKEUCHI [47], THRALL & DUNCAN [48], SCHÜTZENBERGER [41], ROLF [37], and WILLE [53], the following theorem solves the question about finite $\mathfrak{M}$-free closures FM(P) for finite partially ordered sets.

THEOREM 15: [54] For a finite partially ordered set P the following conditions are equivalent:
(i) FM(P) is finite.
(ii) FM(P) is a subdirect product of lattices isomorphic to D_2 or M_3.
(iii) P does not contain an isomorphic copy of $\mathbb{1}+\mathbb{1}+\mathbb{1}+\mathbb{1}$ and $\mathbb{1}+\mathbb{2}+\mathbb{2}$.

Again, condition (ii) gives the possibility to apply the construction methods of our first section for determining the finite lattices FM(P). This was already apparent in the discussion of scaffolding of $FM(\mathbb{1}+\mathbb{1}+\mathbb{n})$. Since the finite partially ordered sets satisfying (iii) are not internally characterized, we can only illustrate condition (iii) by some examples.

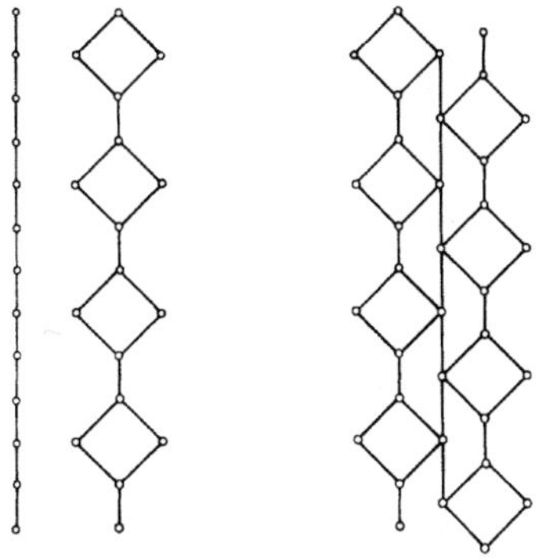

An infinite list of finite simple modular lattices generated by isomorphic copies of the "critical" partially ordered set $\mathbb{1}+\mathbb{1}+\mathbb{1}+\mathbb{1}$ we obtain by the following result (examples for $\mathbb{1}+\mathbb{2}+\mathbb{2}$ can be found in HERRMANN & KINDERMANN & WILLE [20]).

<u>THEOREM 16</u>: (HERRMANN & RINGEL & WILLE [21], HERRMANN [19]) The lattice of all subspaces of an n-dimensional projective geometry over a prime field ($2 \leq n < \infty$) has a generating subset isomorphic to $\mathbb{1}+\mathbb{1}+\mathbb{1}+\mathbb{1}$.

Besides further examples in HERRMANN [19] another infinite list of finite simple modular lattices with four generators is obtained by taking the intervals $[e(n,n),1]$ in $FM(J_1^4)$ which is described in the following theorem; J_1^4 is the partial lattice formed by 0,1, and the atoms of the lattice described by 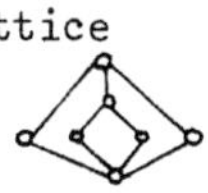.

<u>THEOREM 17</u>: (DAY & HERRMANN & WILLE [11]) The modular lattice $FM(J_1^4)$ is described by the following diagram

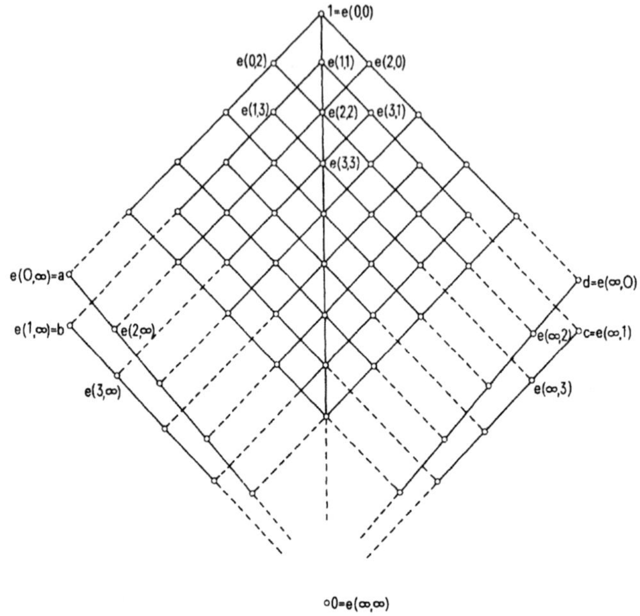

In the proof of Theorem 15 the essential part is to show the implication (iii) ⇒ (ii). This is done with help of the following two lemmas (s. [53]).

D_2-LEMMA: Let M be a subdirectly irreducible modular lattice generated by a finite subset $E_o \cup E_1$. Then $M \not\cong D_2$ implies $\bigvee E_o \geq \bigwedge E_1$.

M_3-LEMMA: Let M be a subdirectly irreducible modular lattice generated by a finite subset $E_o \cup E_2 \cup E_3 \cup E_5 \cup E_1$ $(E_2, E_3, E_5 \neq \emptyset)$; furthermore let $\bar{e}_i := \bigvee \bigcup (E_j \mid i \text{ divides } j)$ and $\underline{e}_i := \bigwedge \bigcup (E_j \mid j \text{ divides } i)$ for $i \in \{2,3,5\}$. Then $M \not\cong M_3$ implies

$$(\bar{e}_2 \wedge \bar{e}_3) \vee (\bar{e}_2 \wedge \bar{e}_5) \vee (\bar{e}_3 \wedge \bar{e}_5) \geq (\underline{e}_2 \vee \underline{e}_3) \wedge (\underline{e}_2 \vee \underline{e}_5) \wedge (\underline{e}_3 \vee \underline{e}_5).$$

For demonstration how to use the lemmas we prove that FM(P) is a subdirect product of lattices isomorphic to D_2 (i.e. FM(P) is distributive) if P is the union of two chains $a_1 \leq a_2 \leq \ldots \leq a_n$ and $b_1 \leq b_2 \leq \ldots \leq b_n$ (c.f. BIRKHOFF [6], p.72). Suppose φ is a homomorphism from P onto a generating subset of a subdirectly irreducible factor M of FM(P) with $M \not\cong D_2$. With $E_o := \{\varphi a_1\}$ and $E_1 := \varphi P \setminus E_o$, we get $\varphi a_1 \geq \varphi a_2 \wedge \varphi b_1$ by the D_2-Lemma; hence $0 = \varphi a_1 \wedge \varphi b_1 = \varphi a_2 \wedge \varphi b_1$ and, by further applications of the D_2-Lemma, $0 = a_2 \wedge \varphi b_1 = \varphi a_3 \wedge \varphi b_1 = \ldots = \varphi a_n \wedge \varphi b_1 = \varphi b_1$. Similarly, it follows $0 = \varphi b_2 = \ldots = \varphi b_n$ and also $0 = \varphi a_1 = \ldots = \varphi a_n$. Thus, we obtain $|M| = 1$ which contradicts the subdirect irreducibility of M. Therefore every subdirectly irreducible factor of FM(P) is isomorphic to D_2.

The D_2-Lemma and the M_3-Lemma have also been used to characterize the finite partial modular lattices $L_o + L_1 + \ldots + L_n$ with finite $\mathcal{M}$-free closure where $L_o + L_1 + \ldots + L_n$ is the disjoint union of the finite modular lattices $(L_i, \vee_i, \wedge_i)$ $(0 \leq i \leq n)$ which means that $x \vee y = z$ in $L_o + L_1 + \ldots + L_n$ if and only if $x,y,z \in L_i$ for some i and $x \vee_i y = z$ and analogously for $\wedge$.

<u>THEOREM 18</u>: [55] For finite modular lattices $L_o, L_1, \ldots, L_n$ $(n \geq 1)$ the following conditions are equivalent:

(i) $FM(L_o + L_1 + \ldots + L_n)$ is finite.

(ii) $FM(L_o + L_1 + \ldots + L_n)$ is a subdirect product of lattices isomorphic to D_2 or M_3.

(iii) Up to a permutation of $L_o, L_1, \ldots, L_n$ either

a) $n = 1$, $|L_o| = 1$, $L_1 \cong S$ where S is a sublattice of the direct product of two chains,

b) $n = 1$, L_o is a chain, $L_1 = \bigcup_{0 \leq i \leq k} [a_i - i, a_i]$ where $[a_{i-1}, a_i] \cong D_2$ or $D_2 \times D_2$, or

c) $n = 2$, $|L_o| = |L_1| = 1$, L_2 is a chain.

Theorem 18 especially answers a question in DORNINGER [14] which asks for the finite modular polynomial lattices over a finite distributive lattice. As a more isolated result for $\mathcal{M}$, we would like to mention the free modular lattices $FM({}_DM_3)$ which is finite for any finite distributive lattice D (s. MITSCHKE & WILLE [31]).

In general, the question about finite $\mathcal{L}$-free closures has only been considered for the class of all lattices and the class of all modular lattices. The only exception is the result that $F(P, \mathcal{L})$ is finite for any finite partial lattice if $\mathcal{L}$ is defined by the set of all lattice equations valid in a fixed finite lattice (s. BIRKHOFF [7], p.144; a slight generalization for modular lattices can be found in [62]). For such a class $\mathcal{L}$ the main problem concerning $F(P, \mathcal{L})$ is very often the question whether the homomorphism $\iota : P \to F(P, \mathcal{L})$ is injective or not (e.g. [51], Satz 3.4). An example showing that the study of $\mathcal{L}$-free closures for a specific class $\mathcal{L}$ can be very interesting is given in [57] where the existence problem for the projective plane of order 10 is turned into a question whether a certain $\mathcal{L}$-free closure has more than 16 elements.

REFERENCES

1. B. Banaschewski: Hüllensysteme und Erweiterungen von Quasi-Ordnungen. Z.Math.Logik Grundlagen Math. <u>2</u> (1956), 117-130.

2. G. Birkhoff: On the combination of subalgebras. Proc. Cam-

bridge Phil. Soc. 29 (1933), 441-464.

3. G. Birkhoff: On the structure of abstract algebras. Proc. Cambridge Phil. Soc. 31 (1935), 433-454.
4. G. Birkhoff: Rings of sets. Duke Math.J. 3 (1937), 442-454.
5. G. Birkhoff: Subdirect unions in universal algebras. Bull. Amer.Math.Soc. 50 (1944), 764-768.
6. G. Birkhoff: Lattice theory. 2nd ed. Providence: Amer.Math. Soc. 1948.
7. G. Birkhoff: Lattice theory. 3rd ed. Providence: Amer.Math. Soc. 1967.
8. H.H. Crapo, G.-C. Rota: On the foundations of combinatorial theory: Combinatorial geometries (preliminary edition). Cambridge-London: M.I.T. Press 1970.
9. P. Crawley: Decomposition theory of nonsemimodular lattices. Trans.Amer.Math.Soc. 99 (1961), 246-254.
10. P. Crawley, R.P. Dilworth: Algebraic theory of lattices. Englewood Cliffs: Prentice Hall 1973.
11. A. Day, C. Herrmann, R. Wille: On modular lattices with four generators. Alg. Universalis 2 (1972), 317-323.
12. R. Dedekind: Über die von drei Moduln erzeugte Dualgruppe. Math.Ann. 53 (1900), 371-403.
13. R.P. Dilworth: Lattices with unique irreducible decompositions. Ann.Math. 41 (1940), 771-777.
14. D. Dorninger: Modulare Polynomverbände über endlichen distributiven Verbänden. Monatsh.Math. 78 (1974), 305-310.
15. D. Dorninger, J. Wiesenbauer: Anzahlsätze für Polynomfunktionen auf Verbänden. Preprint.
16. B. Ganter, W. Poguntke, R. Wille: Finite sublattices of four-generated modular lattices. Preprint.
17. G. Gierz, K. Keimel: Topologische Darstellung von Verbänden. Math.Z. 150 (1976), 83-99.
18. G. Grätzer: Lattice theory: First concepts and distributive lattices. San Francisco: Freeman 1971.
19. C. Herrmann: On the equational theory of submodule lattices. In: Proc. Lattice Theory Conf. Houston 1973, pp. 105-118.
20. C. Herrmann, M. Kindermann, R. Wille: On modular lattices generated by $\mathbb{1}+\mathbb{2}+\mathbb{2}$. Alg. Universalis 5 (1975), 243-251.
21. C. Herrmann, C.M. Ringel, R. Wille: On modular lattices with four generators. Not.Amer.Math.Soc. 20 (1973), A-418; 73T-A151.
22. B. Jónsson: Algebras whose congruence lattices are distributive. Math. Scand. 21 (1967), 110-121.
23. B. Jónsson, J.B. Nation: A report on sublattices of a free lattice. In: Contributions to Universal Algebra. Szeged: J. Bolyai Mat. Soc. (to appear).
24. D. Kelly: The 3-irreducible partially ordered sets. Canad. J.Math. (submitted).
25. F. Klein: Über einen Zerlegungssatz in der Theorie der abstrakten Verknüpfungen. Math. Ann. 106 (1932), 114-130.
26. A. Kurosch: Durchschnittsdarstellungen mit irreduziblen Komponenten in Ringen u. sogen. Dualgruppen. Mat.Sb. 42(1935),613-616.

27. H. Lausch, W. Nöbauer: Algebra of polynomials. Amsterdam-London: North-Holland 1973.

28. G. Markowsky: Combinatorial aspects of lattice theory with applications to the enumeration of free distributive lattices. Ph.D.Thesis, Harvard University 1973.

29. G. Markowsky: Some combinatorial aspects of lattice theory. In: Proc. Lattice Theory Conf. Houston 1973, pp. 36-68.

30. R. McKenzie: Equational bases and non-modular lattice varieties. Trans. Amer. Math. Soc. 174 (1972), 1-43.

31. A. Mitschke, R. Wille: Freie modulare Verbände $FM({}_DM_3)$. In: Proc. Lattice Theory Conf. Houston 1973, pp. 383-396.

32. O. Ore: On the foundation of abstract algebra II. Ann.Math. 37 (1936), 265-292.

33. H.A. Priestley: Representation of distributive lattices by means of ordered Stone spaces. Bull. London Math. Soc. 2 (1970), 186-190.

34. H.A. Priestley: Ordered topological spaces and the representation of distributive lattices. Proc. London Math. Soc. III. Ser. 24 (1972), 507-530.

35. I. Rival: Contributions to combinatorial lattice theory. Ph.D. Thesis, Winnipeg 1973.

36. H.L. Rolf: The free lattice generated by a set of chains. Pacific J. Math. 8 (1958), 585-595.

37. H.L. Rolf: The free modular lattice, $FM(2+2+2)$, is infinite. Proc.Amer. Math.Soc. 17 (1966), 960-961.

38. E.T. Schmidt: Kongruenzrelationen algebraischer Strukturen. Math.Forschungsber. 25, Berlin: VEB Dt. Verlag d. Wiss. 1969.

39. J. Schmidt: Zur Kennzeichnung der Dedekind-MacNeillschen Hülle einer geordneten Menge. Arch.Math. 7 (1956), 241-249.

40. J. Schmidt: Boolean duality extended. In: Theory of sets and topology. A collection of papers in honour of Felix Hausdorff. Berlin: VEB Dt. Verlag d. Wiss. 1972, pp. 435-453.

41. M.P. Schützenberger: Construction du treillis modulaire engendré par deux éléments et une chaîne finie discrète. C.R. Acad. Sci. Paris 235 (1952), 926-928.

42. D. Schweigert: Über endliche ordnungspolynomvollständige Verbände. Monatsh. Math. 78 (1974), 68-76.

43. Yu.I. Sorkin: Free unions of lattices. Mat. Sb. 30 (72) (1952), 677-694 (russian).

44. M.H. Stone: The theory of representations of Boolean algebras. Trans.Amer.Math.Soc. 122 (1936), 379-398.

45. H. Strietz: Finite partition lattices are four-generated. In: Proc. Lattice Theory Conf. Ulm 1975, 257-259.

46. H. Strietz: Über Erzeugendensysteme endlicher Partitionsverbände. Preprint.

47. K. Takeuchi: On free modular lattices II. Tôhoku Math.J. 11 (1959), 1-12.

48. R.M. Thrall, D.G. Duncan: Note on free modular lattices. Amer.J.Math. 75 (1953), 627-632.

49. A. Urquhart: A topological representation theory for lattices.

Preprint.

50. A.G. Waterman: The free lattice with 3 generators over N_5. Port.Math. 26 (1967), 285-288.
51. R. Wille: Primitive Länge und primitive Weite bei modularen Verbänden. Math.Z. 108 (1969), 129-136.
52. R. Wille: Subdirekte Produkte und konjunkte Summen. J. reine angew. Math. 239/240 (1970), 333-338.
53. R. Wille: On free modular lattices generated by finite chains. Alg. Universalis 3 (1973), 131-138.
54. R. Wille: Über modulare Verbände, die von einer endlichen halbgeordneten Menge frei erzeugt werden. Math.Z. 131 (1973), 241-249.
55. R. Wille: Über freie Produkte endlicher modularer Verbände. Abh. Math. Sem. Hamburg 45 (1976), 218-224.
56. R. Wille: Subdirekte Produkte vollständiger Verbände. J. reine angew. Math. 283/284 (1976), 53-70.
57. R. Wille: Finite projective planes and equational classes of modular lattices. In: Atti. Conv. Geom. Comb. Roma 1973, (to appear).
58. R. Wille: A note on simple lattices. In: Contributions to Lattice Theory. Szeged: J. Bolyai Mat. Soc. (to appear).
59. R. Wille: On lattices freely generated by finite partially ordered sets. In: Contribution to Universal Algebra. Szeged: J. Bolyai Mat. Soc. (to appear).
60. R. Wille: Eine Charakterisierung endlicher, ordnungspolynomvollständiger Verbände. Arch. Math. (to appear).
61. R. Wille: A note on algebraic operations and algebraic functions of finite lattices. Preprint.
62. R. Wille: Jeder endlich erzeugte, modulare Verband endlicher Weite ist endlich. Math. čas. 24 (1974), 77-80.

PART III

MATROIDS

GEOMETRIES ASSOCIATED WITH PARTIALLY ORDERED SETS

T. A. Dowling* and W. A. Denig

Department of Mathematics, The Ohio State University, Columbus, Ohio 43210

ABSTRACT. The subsets of a finite partially ordered set P which occur as the tops of chains of length $\geq k$ in some k- and $(k-1)$-saturated chain partition of P are shown to be the bases of a combinatorial geometry $G_k(P)$, thus affirming a conjecture of Greene. The identity map on P is a strong map $G_{k-1}(P) \to G_k(P)$ for $k = 1, 2, \ldots$, and is representable as a linear map over every sufficiently large field. $G_k(P)$ is a gammoid, but every gammoid is a subgeometry of $G_2(P)$ for some partially ordered set P of height 3.

0. INTRODUCTION

In a recent paper [7], Greene and Kleitman introducted the extremal concepts of Sperner k-families and k-saturated chain partitions in a a finite partially ordered set P and proved several beautiful theorems about them. Their results led Greene to conjecture in a later paper [8] that certain subsets of P, associated with k-saturated chain partitions, are the independent

*Research supported by National Science Foundation Grant No. MC576-10042 (Ohio State University Research Foundation Project No. 4451-A1).

M. Aigner (ed.), Higher Combinatorics, 103-116. *All Rights Reserved.*

sets of a combinatorial geometry[1] [2] on P. A proof of Greene's conjecture appears in [4], together with other results on the structure and properties of the geometries which arise in this way. Several of these results from [4] are given in this article. Proofs are only outlined, or in some cases, omitted.

1. SPERNER k-FAMILIES AND SATURATED CHAIN PARTITIONS

This section is devoted primarily to a review of definitions and theorems which appear in [7].

A k-family in a finite partially ordered set P is a subset of P which contains no chain of length (size) $k + 1$, or equivalently, a subset of P which can be expressed as the union of k antichains. The maximum size of a k-family in P is denoted by $d_k(P)$, and a k-family of size $d_k(P)$ is called a Sperner k-family. The basic relationship between 1-families (antichains) and chain partitions of P was established in a fundamental theorem of Dilworth [7]:

Theorem 1. A finite partially ordered set P can be partitioned into $d_1(P)$ chains.

Any chain partition $\mathcal{C}$ of P induces an upper bound on the size of a k-family, and hence on $d_k(P)$, as follows. Since a k-family A can meet a chain C in at most $\min\{k, |C|\}$ elements,

$$|A| = \sum_{C \in \mathcal{C}} |A \cap C| \leq \beta_k(\mathcal{C})$$

where

$$\beta_k(\mathcal{C}) = \sum_{C \in \mathcal{C}} \min\{k, |C|\} .$$

Thus

$$d_k(P) \leq \beta_k(\mathcal{C})$$

for any chain partition $\mathcal{C}$ of P. $\mathcal{C}$ is called k-saturated if equality holds. Since $\beta_1(\mathcal{C})$ is just the number of chains in $\mathcal{C}$, Theorem 1 asserts that 1-saturated chain partitions exist.

[1] We use the term "geometry" in place of "pregeometry" or "matroid."

Theorem 1 was generalized to arbitrary k by Greene and Kleitman in [7]:

Theorem 2. For any $k \geq 0$, a k-saturated chain partition exists.

Given a chain partition $\mathcal{C}$, let

$$\mathcal{C}_k = \{C \in \mathcal{C} : |C| \geq k\}$$

and define

$$\alpha_k(\mathcal{C}) = \sum_{C \in \mathcal{C}_k} (|C| - k) .$$

We then have

$$\alpha_k(\mathcal{C}) + \beta_k(\mathcal{C}) = |P| ,$$

and therefore

$$\alpha_k(\mathcal{C}) \leq |P| - d_k(P)$$

with equality holding if and only if $\mathcal{C}$ is k-saturated. Clearly the k-saturated property of a chain partition depends only upon its set $\mathcal{C}_k$ of chains of length $\geq k$.

The integers $\Delta_k(P)$ are defined by

$$\Delta_k(P) = d_k(P) - d_{k-1}(P) .$$

Then, since

$$\alpha_{k-1}(\mathcal{C}) = \alpha_k(\mathcal{C}) + |\mathcal{C}_k| ,$$

a k-saturated chain partition $\mathcal{C}$ satisfies

$$|\mathcal{C}_k| = d_k(P) - \beta_{k-1}(\mathcal{C}) ,$$

and thus

$$|\mathcal{C}_k| \leq \Delta_k(P) ,$$

with equality holding if and only if $\mathcal{C}$ is also $(k-1)$-saturated. We shall call such a partition a $(k,k-1)$-saturated chain partition. Again it is clear that this property also depends only on $\mathcal{C}_k$. That such partitions exist was proved in [7]:

Theorem 3. For any $k \geq 1$, a $(k,k-1)$-saturated chain partition exists.

This theorem is best possible in the sense that examples show there may be no chain partition which is k-saturated for three consecutive values of k.

Other examples in [7] demonstrate that a Sperner $(k-1)$-family A need not be contained in a Sperner k-family, and thus the inequality $\Delta_k(P) \geq d_1(P - A)$ may hold strictly. The inequalities of the following theorem [7] are therefore not immediate. (The height of a partially ordered set is the length of its longest chain.)

Theorem 4. Let P be of height ℓ. Then

$$\Delta_1 \geq \Delta_2 \geq \cdots \geq \Delta_\ell > 0,$$

and

$$\Delta_k = 0 \quad \text{for } k > \ell .$$

2. GREENE'S CONJECTURE

Given a chain partition $\mathcal{C}$ of P, let Top $\mathcal{C}_k$ denote the set of top elements of the chains of $\mathcal{C}_k$. The following theorem was proved by Greene and Kleitman in [7]:

Theorem 5. Let X be the set of maximal elements of a finite partially ordered set Q, and let $P = Q - X$. Then the set function r_k defined on subsets of X by

$$r_k(A) = |A| - \delta_k(A),$$

where

$$\delta_k(A) = d_k(A \cup P) - d_k(P),$$

is the rank function of a combinatorial geometry on X. A subset of X is independent if and only if it can be matched downward in Q into Top $\mathcal{C}_k$, for some k-saturated chain partition $\mathcal{C}$ of P.

For any sets X and Y, a relation $R \subseteq X \times Y$ and a geometry H on Y induces a geometry $G = R^{-1}H$ on X whose independent sets are those subsets of X which can be matched under R to some independent set of $H(Y)$. (See e.g. [2] or [10].)

Theorem 5 quite naturally led Greene [8] to make the

Conjecture. Those subsets $I \subseteq P$, for which $I \subseteq \text{Top}\, \mathcal{C}_k$ for some k-saturated chain partition $\mathcal{C}$, are the independent sets of a combinatorial geometry on P.

The conjecture implies that the geometry of Theorem 5 exists as an induced geometry under the relation $R \subseteq X \times P$ given by xRp if $x > p$ in Q.

3. k-LINKINGS AND CHAIN PARTITIONS

Let Γ be a finite directed graph. Given any set $\mathcal{A}$ of vertex-disjoint paths in Γ, denote by $\text{In}\,\mathcal{A}$ the set of initial vertices and by $\text{Ter}\,\mathcal{A}$ the set of terminal vertices of paths in $\mathcal{A}$. If X, Y are subsets of the vertex set V of Γ, an (X,Y)-linking in Γ is a set $\mathcal{A}$ of vertex-disjoint paths in Γ for which $\text{In}\,\mathcal{A} \subseteq X$ and $\text{Ter}\,\mathcal{A} \subseteq Y$. Mason [10] showed that the subsets $\text{In}\,\mathcal{A}$, for $\mathcal{A}$ an (X,Y)-linking in Γ, are the independent sets of a combinatorial geometry $G = (\Gamma; X,Y)$ on X, called a gammoid. The geometry $(\Gamma; V,Y)$ is a strict gammoid. Y is the distinguished basis in this presentation of G. Any basis can be taken as the distinguished basis relative to some directed graph on V. If F is a closed set of the strict gammoid $(\Gamma; V,Y)$, then the subset $F_0 = \{x \in F: \Gamma x \not\subseteq F\}$ is a basis of F, where Γx is the set of vertices joined from x by an edge of Γ.

Let $P_0, P_1, \ldots, P_k$ be $k+1$ disjoint copies of a finite partially ordered set P, and denote the image of $x \in P$ by $x_i \in P_i$ under the canonical bijection, and of $A \subseteq P$ by $A_i \subseteq P_i$. Following Greene [8], let us define a directed graph $\Gamma_k(P)$ with vertex set $V = P_0 \cup P_1 \cup \ldots \cup P_k$ and edge set

$$E = \{(x_{i-1}, y_i): x > y, \quad 1 \leq i \leq k\} .$$

A (P_0, P_k)-linking in $\Gamma_k(P)$ will be called a k-linking. The maximum number of paths in a k-linking is denoted $\mu_k(P)$.

To a chain partition $\mathcal{C}$ of P, we may associate a k-linking $\mathcal{A} = \mathcal{A}(\mathcal{C}_k)$ whose paths are the segments of length $k+1$ in chains of $\mathcal{C}_k$. The size of $\mathcal{A}$ is $\alpha_k(\mathcal{C})$, so $\mu_k(P) \geq |P| - d_k(P)$, since $\alpha_k(\mathcal{C}) = |P| - d_k(P)$ when $\mathcal{C}$ is k-saturated. The following theorem is stated in [8].

Theorem 6. The maximum size of a k-linking in $\Gamma_k(P)$ is $\mu_k(P) = |P| - d_k(P)$.

Outline of proof. If A is a Sperner k-family, and $x \in P - A$, there is no path from A_0 to A_k in the subgraph $\Gamma_k(A)$ of $\Gamma_k(P)$, but there is a path from $A_0 \cup \{x_0\}$ to $A_0 \cup \{x_k\}$ in $\Gamma_k(A \cup x)$. Every such path must contain x_i for some uniquely determined i, depending on x. The set S consisting of these x_i's can be shown to separate P_0 from P_k in $\Gamma_k(P)$, and therefore

$$\mu_k(P) \leq |S| = |P| - d_k(P).$$

Fulkerson [6] observed that the edges of a maximum 1-linking (matching) in $\Gamma_1(P)$ can be joined to form a minimum chain partition of P, and applied König's Theorem on bipartite graphs to obtain another proof of Dilworth's Theorem. Under this correspondence between maximum matchings $\mathcal{A}$ and minimum chain partitions $\mathcal{C}$, $P_1 \setminus \mathrm{Ter}\,\mathcal{A} = (\mathrm{Top}\,\mathcal{C})_1$. Since $\mathrm{Ter}\,\mathcal{A}$ is a basis of the transversal geometry on P_1 presented by $\Gamma_1(P)$, we may state

Theorem 7 [6, 8]. The subsets $\mathrm{Top}\,\mathcal{C}$, for $\mathcal{C}$ a minimum chain partition of P, are the bases of a dual transversal geometry $G_1(P)$ on P.

Denig [4] has a constructive characterization of those dual transversal geometries $G(X)$ which occur as $G_1(X)$ for some partial order on X.

4. (k, k - 1)-LINKINGS AND CHAIN PARTITIONS

To extend Fulherson's argument from the case $k = 1$, we view the set $P_1 \setminus \mathrm{Ter}\,\mathcal{A}$ in $\Gamma_1(P)$ as the initial vertices of a set $\mathcal{B}$ of paths of length $k = 1$ from P_1 to P_k, which are disjoint from paths in $\mathcal{A}$. In general, we define a $(k, k-1)$-linking in $\Gamma_k(P)$ as a pair $(\mathcal{A}, \mathcal{B})$ such that

(1) $\mathcal{A}$ is a (P_0, P_k)-linking,

(2) $\mathcal{B}$ is a (P_1, P_k)-linking,

(3) $\mathcal{A} \cup \mathcal{B}$ is a $(P_0 \cup P_1, P_k)$-linking.

The size of $(\mathcal{A}, \mathcal{B})$ is the pair $(|\mathcal{A}|, |\mathcal{B}|)$. We shall consider $(\mathcal{A}, \mathcal{B})$ to be a maximum $(k, k-1)$-linking if, for any other $(k, k-1)$-linking $(\mathcal{A}', \mathcal{B}')$, we have $|\mathcal{A}| \geq |\mathcal{A}'|$ and $|\mathcal{A} \cup \mathcal{B}| \geq |\mathcal{A}' \cup \mathcal{B}'|$. If $\mathcal{A}^T$ denotes the (P_1, P_k)-linking obtained by deleting the initial vertices of paths in $\mathcal{A}$, then $\mathcal{A}^T \cup \mathcal{B}$ in a $(k-1)$-linking in the induced subgraph on $P_1 \cup P_2 \cup \cdots \cup P_k$ isomorphic to $\Gamma_{k-1}(P)$. Thus $|\mathcal{A} \cup \mathcal{B}| = |\mathcal{A}^T \cup \mathcal{B}| \leq \mu_{k-1}(P) = |P| - d_{k-1}(P)$. Thus for any maximum $(k, k-1)$-linking, $|\mathcal{A}| = |P| - d_k(P)$ and $|\mathcal{B}| \leq \Delta_k(P)$.

Given a chain partition $\mathcal{C}$ of P, let $\mathcal{B} = \mathcal{B}(\mathcal{C}_k)$ be the set of paths from P_1 to P_k in $\Gamma_k(P)$ corresponding to the top segments of length k in chains of $\mathcal{C}_k$. Then if $\mathcal{C}$ is k-saturated, $|\mathcal{B}| = |\mathcal{C}_k| \leq \Delta_k(P)$, with equality holding when $\mathcal{C}$ is $(k-1)$-saturated. Taking $\mathcal{C}$ to be $(k, k-1)$-saturated, we obtain

Theorem 8. The size of a maximum $(k, k-1)$-linking in $\Gamma_k(P)$ is $(|P| - d_k(P), \Delta_k(P))$.

Although a $(k, k-1)$-saturated chain partition $\mathcal{C}$ yields a maximum $(k, k-1)$-linking $(\mathcal{A}_k(\mathcal{C}), \mathcal{B}(\mathcal{C}))$ in $\Gamma_k(P)$, not all maximum linkings arise in this way when $k > 1$. We require only a partial converse, however. The following theorem is the main step in our proof of Greene's conjecture. Accordingly, a rather detailed outline of the proof is given.

Theorem 9. Let $(\mathcal{A}, \mathcal{B})$ be a maximum $(k, k-1)$-linking in $\Gamma_k(P)$. Then there is a $(k, k-1)$-saturated chain partition $\mathcal{C}$ of P such that $(\mathrm{Top}\, \mathcal{C}_k)_1 = \mathrm{In}\, \mathcal{B}$.

Outline of proof. We argue by induction on $|P|$, considering two cases. In Case 1, some maximal element b of P satisfies $d_k(P - b) = d_k(P)$, and we appeal to a result of Greene and Kleitman ([7], Theorem 3.10, Statement (1) in the proof) to conclude that $d_{k-1}(P \setminus b) = d_{k-1}(P)$. Since $\mu_k(P \setminus b) = \mu_k(P) - 1$, $b_0 \in \mathrm{In}\, \mathcal{A}$, and we let a_1 be the second vertex on the $\mathcal{A}$-path A containing b_0. We next show that $b_1 \in \text{In}\,\mathcal{B}$, and let B be the $\mathcal{B}$-path containing b_1. On removing A from $\mathcal{A}$ and replacing B by A^T in $\mathcal{B}$, we obtain a maximum $(k,k-1)$-linking in $\Gamma_k(P\setminus b)$. The inductive hypothesis gives a $(k,k-1)$-saturated chain partition $\mathcal{D}$ of $P\setminus\{b\}$ with $(\text{Top}\,\mathcal{D}_k)_1 = (\text{In}\,\mathcal{B}\setminus\{b_1\}) \cup \{a_1\}$. Adjoining b to the chain of $\mathcal{D}_k$ with top a gives the required chain partition of P.

In Case 2, every maximal element b of P satisfies $d_k(P\setminus b) = d_k(P) - 1$, which implies $\Delta_k(P) \geq \Delta_k(P\setminus b)$. We consider an element $b \in \text{Top}\,\mathcal{C}'_k$ for some $(k,k-1)$-saturated chain partition $\mathcal{C}'$, assume that $b_1 \notin \text{In}\,\mathcal{B}$, and show that b is maximal in P. Hence no path of $\mathcal{B}$ contains a b_i. The next step is to show that there exists a maximum $(k,k-1)$-linking $(\mathcal{A}',\mathcal{B})$ such that $b_0 \notin \text{In}\,\mathcal{A}'$, so that no b_i is on a path of $\mathcal{A}'$. The argument is as follows. Since $\mu_k(P\setminus b) = \mu_k(P)$, we can find a maximum k-linking $\mathcal{A}''$ such that $b_0 \notin \text{In}\,\mathcal{A}''$. The closure F of P_0 in the strict gammoid $(\Gamma_k;\, V, P_k)$ cannot contain a vertex on a path of $\mathcal{B}$, and each path of $\mathcal{A}''$ meets a path of $\mathcal{A}$ in a unique vertex of the F-basis $F_0 = \{x \in F\colon \Gamma x \not\subseteq F\}$. We construct each path of $\mathcal{A}'$ by joining the initial segment of a path in $\mathcal{A}''$, preceding the vertex of F_0, to the terminal segment of the path of $\mathcal{A}$ which contains that F_0-vertex.

Then $(\mathcal{A}',\mathcal{B})$ is a maximum $(k,k-1)$-linking in $P\setminus\{b\}$, and by the inductive hypothesis there is a $(k,k-1)$-saturated chain partition $\mathcal{D}$ of $P\setminus\{b\}$ such that $(\text{Top}\,\mathcal{D}_k)_1 = \text{in}\,\mathcal{B}$. Then $\mathcal{C} = \mathcal{D} \cup \{b\}$ satisfies the theorem.

5. THE GEOMETRY $G_k(P)$

Consider now the strict gammoid $G = (\Gamma_k;\, V, P_k)$ on $V = P_0 \cup P_1 \cup \dots \cup P_k$. The bases of the restriction $G(P_0)$ are the sets $\text{In}\,\mathcal{A}$, for $\mathcal{A}$ maximum k-linking in $\Gamma_k(P)$. The bases of the restriction $G(P_0 \cup P_1)$ are the sets $\text{In}\,\mathcal{A} \cup \text{In}\,\mathcal{B}$, for $(\mathcal{A},\mathcal{B})$ a $(k,k-1)$-linking in $\Gamma_k(P)$ with $|\mathcal{A} \cup \mathcal{B}| = |P| - d_{k-1}(P)$. Thus the bases of the minor $G(P_0 \cup P_1)/P_0$ are the sets $\mathrm{In}\,\mathcal{B}$, for $(\mathcal{A},\mathcal{B})$ a maximum $(k,k-1)$-linking in $\Gamma_k(P)$. Since a minor of a gammoid is a gammoid [9], we may state

<u>Theorem 10</u>. The subsets $\mathrm{Top}\,\mathcal{C}_k$, for $\mathcal{C}$ a $(k,k-1)$-saturated chain partition of P, are the bases of a combinatorial geometry $G_k(P)$ on P. This geometry is a gammoid of rank $\Delta_k(P)$.

That this geometry is identical to the one Greene conjectured now follows from

<u>Theorem 11</u>. If $\mathcal{C}$ is a k-saturated chain partition of P, there exists a $(k,k-1)$-saturated chain partition $\mathcal{C}'$ such that $\mathrm{Top}\,\mathcal{C}_k \subseteq \mathrm{Top}\,\mathcal{C}'_k$,

<u>Proof</u>. Let $(\mathcal{A},\mathcal{B})$ be the $(k,k-1)$-linking associated with $\mathcal{C}$. Then $\mathrm{In}\,\mathcal{A} \cup \mathrm{In}\,\mathcal{B}$ is independent in $G(P_0 \cup P_1)$, and since $\mathcal{C}$ is k-saturated, $\mathrm{In}\,\mathcal{A}$ is a basis of $G(P_0)$. Thus $\mathrm{In}\,\mathcal{B}$ is independent in $G(P_0 \cup P_1)/P_0$, and so is contained in a basis $\mathrm{In}\,\mathcal{B}'$. But $\mathrm{In}\,\mathcal{B} = (\mathrm{Top}\,\mathcal{C}_k)_1$ and $\mathrm{In}\,\mathcal{B}' = (\mathrm{Top}\,\mathcal{C}'_k)_1$ for some $(k,k-1)$-saturated chain partition $\mathcal{C}'$ of P.

<u>Corollary</u>. Greene's conjecture is true.

6. THE STRONG MAP $G_{k-1}(P) \to G_k(P)$

If G and H are combinatorial geometries on the same set X and the identity map on X is a <u>strong map</u> from G to H, i.e. if every H-closed set is G-closed, we say $G \to H$ is a strong map. A strong map $G \to H$ is <u>elementary</u> if $r(G) = r(H) + 1$, where r denotes rank. An elementary strong map $G \to H$ is characterized uniquely by an extension [2] of G by a single point p, whose contraction yields H. An elementary strong map is <u>principal</u> [5] if the modular cut defining the extension is a principal order filter in the lattice of closed sets of G, i.e. if it has a unique generating flat. If F is such a flat, then the rank functions of G and H are related by $r_H(A) = r_G(A)$ or $r_G(A) - 1$ according as the G-closure of A does not contain F or contains F.

Let $F(X)$ denote the free geometry on X, in which all subsets are closed. Then for any geometry $G(X)$, $F(X) \to G(X)$ is a strong map, the <u>closure map</u> of G.

The following theorem is due to Dowling and Kelly [5], and Brown [1].

> Theorem 12. A geometry $G(X)$ is a dual transversal geometry if and only if the closure map $F(X) \to G(X)$ can be factored into principal strong maps.

The proof shows that the G-closures of the generating flats in a principal factorization give the unique maximal presentation of the transversal geometry $G^*(X)$.

A second characterization of dual transversal geometries was given by Ingleton and Piff [9]:

> Theorem 13. A geometry $G(X)$ is a dual transversal geometry if and only if it is a strict gammoid.

The following theorem displays the connection between these two characterizations of dual transversal geometries.

> Theorem 14. Let Y be a subset of the vertex set V of a digraph Γ and let $y \in Y$. Then
>
> $$(\Gamma;\ V,Y) \to (\Gamma;\ V,Y - y)$$
>
> is a principal strong map between strict gammoids, generated by the closure of $\Gamma y \cup \{y\}$.

We omit the proof of this theorem. There are some interesting corollaries.

> Corollary 1. Let X, Y, Z be subsets of the vertex set V of a digraph Γ, and suppose every path from X to Z meets Y. Then
>
> $$(\Gamma;\ X,Y) \to (\Gamma;\ X,Z)$$
>
> is a strong map between gammoids.

Proof. By repeated application of Theorem 14 we have that

$$(\Gamma;\ V,Y \cup Z) \to (\Gamma;\ V,Z)$$

is strong, and hence the restriction

$$(\Gamma;\ X,Y \cup Z) \to (\Gamma;\ X,Z)$$

is strong. But a subset $A \subseteq X$ links into $Y \cup Z$ if and only if X links into Y, so $(\Gamma;\ X,Y \cup Z) = (\Gamma;\ X,Y)$.

Corollary 2. For each $k \geq 1$,

$G_{k-1}(P) \to G_k(P)$

is a strong map.

Proof. $G_0(P)$ is free, so the result holds for $k = 1$. Let $k \geq 2$, $V = P_0 \cup P_1 \cup \dots \cup P_k$, and consider the strict gammoids $G = (\Gamma_k; V, P_{k-1} \cup P_k)$ and $H = (\Gamma_k; V, P_k)$. By Theorem 14, $G \to H$ is a strong map, and hence

$G(P_0 \cup P_1)/P_0 \to H(P_0 \cup P_1)/P_0$

is a strong map between minors. But $G(P_0 \cup P_1) = (\Gamma_k; P_0 \cup P_1, P_{k-1} \cup P_k) = (\Gamma_k; P_0 \cup P_1, P_{k-1}) = (\Gamma_{k-1}; P_0 \cup P_1, P_{k-1})$ by Corollary 1, and thus $G(P_0 \cup P_1)/P_0 = G_{k-1}(P)$, while $H(P_0 \cup P_1)/P_0 = G_k(P)$.

Corollary 3 (cf. Theorem 4). Let ℓ be the height of P. Then $\Delta_1 \geq \Delta_2 \geq \dots \geq \Delta_\ell > 0$, and $G_{k-1}(P) = G_k(P)$ if and only if $\Delta_{k-1} = \Delta_k$.

Proof. If $G \to H$ is a strong map, then $r(G) \geq r(H)$ with equality if and only if $G = H$ [2]. The corollary now follows since $r(G_k) = \Delta_k$.

Thus to every finite poset P we can associate a strong sequence $(G_1, G_2, \dots, G_\ell)$ of geometries on P, where ℓ is the height of P, such that $G_i \to G_j$ is a strong map whenever $i \leq j$, and $r(G_i) = \Delta_i$. It is natural to ask how much information about the partial order P is preserved in the strong sequence associated with P. We have only partial answers to this question at present [4]. The following example illustrates that the strong sequence does not determine the partial order.

Example. Two partial orders with the same strong sequence. The Δ-sequence is $(5,3,2,2,1,1)$.

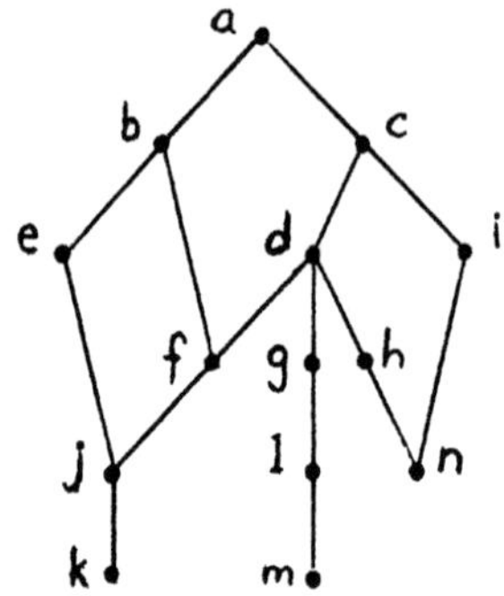

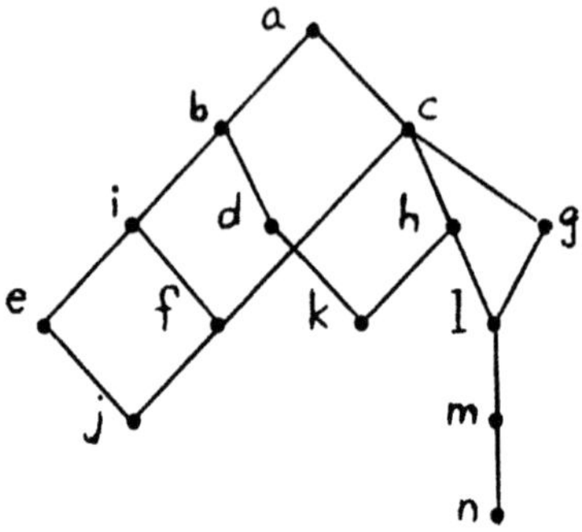

7. LINEAR STRONG MAPS

A representation of a geometry $G(X)$ over a field K is a mapping $\sigma: X \to K^{r(G)}$ such that $I \subseteq X$ is independent in G if and only if $\sigma|I$ is injective and $\sigma(I)$ is linearly independent in $K^{r(G)}$. G is linear over K if a representation of G over K exists.

A representation of a strong map $G \to H$ over K is a linear map $L: K^{r(G)} \to K^{r(H)}$ such that $L \circ \sigma$ is a representation of H over K for any representation σ of G over K. Such maps can exist, of course, only when both G and H are linear over K. We say that $G \to H$ is linear over K if a representation exists.

Theorem 15. If $G \to H$ is a principal strong map, G is linear over K, and the order of K is sufficiently large, then $G \to H$ is linear over K. In particular, $G \to H$ is linear over some finite extension of K.

Outline of proof. Let $t = r(F)$, $n = r(G)$. We choose a basis $b_1, b_2, \ldots, b_t$ of F and extend it to a basis $b_1, b_2, \ldots, b_n$ of G. Let σ be a representation of G with $\sigma(b_i) = \underline{e}_i$, the i-th unit vector of the standard basis of K^n. Since X is finite, there is, for K sufficiently large, a vector $\underline{u} = (u_1, u_2, \ldots, u_t, 0, \ldots, 0)$ in the linear span of $\sigma(F)$ which is not in the linear span of $\sigma(F) \cap \sigma(C)$, for any hyperplane C of G not containing F. Thus $u_i \neq 0$ for $i = 1,2,\ldots,t$, and we can assume $u_1 = 1$. Define $L: K^n \to K^{n-1}$ by $L(\underline{e}_1) = (u_2,\ldots,u_t,0,\ldots,0)$, and $L(\underline{e}_i) = \underline{e}'_{i-1}$, $2 \leq i \leq n$, where the $\underline{e}'_i$ are the standard basis vectors of K^{n-1}. That $L \circ \sigma$ is a representation of H can then be proved using the Cauchy-Binet Theorem.

Corollary. The strong map

$$G_{k-1}(P) \to G_k(P)$$

is linear over every sufficiently large field K.

Proof. It is known [10] that a gammoid is linear over every sufficiently large field. Let G and H be as in the proof of Corollary 2 of Theorem 14. Then the strong map $G \to H$ factors into a product of principal strong maps by Theorem 14, and hence is linear over a sufficiently large field K, by Theorem 15.

The restriction to minors of a linear map is linear, and thus

$$G_{k-1}(P) = G(P_0 \cup P_1)/P_0 \to H(P_0 \cup P_1)/P_0 = G_k(P)$$

is linear.

8. GAMMOIDS IN $G_2(P)$

Minors of gammoids are gammoids [9], so the geometries $G_k(P)$ are gammoids. The following result is therefore somewhat surprising:

Theorem 16. Let G(X) be a gammoid. Then there exists a set S disjoint from X and a partial order on $P = X \cup S$ such that

$$G_2(P) = G(X) \oplus H(S),$$

where H(S) consists entirely of loops and isthmuses.

Proof. By a theorem of Ingleton and Piff [9], every gammoid is a contraction of a transversal geometry. Thus $G(X) = H(X \cup Y)/Y$ for some transversal geometry $H(X \cup Y)$. We can assume that Y is independent in H, and let $R \subseteq (X \cup Y) \times Z$ be a presentation of H, where $|Z| = r(H)$. Let W be disjoint from X, Y, and Z with $|W| = |Y|$, and let $\sigma: W \to Y$ be a bijection. We define a partial order on $P = X \cup (Y \cup Z \cup W)$ by saying $a > b$ if one of the following holds:

(1) $a \in X \cup Y$, $b \in Z$, aRb.

(2) $a \in W$, $b \in Y$, $\sigma(a) = b$.

(3) $a \in W$, $b \in Z$, $\sigma(a)Rb$.

P is then a partial order of height 3.

It is then not difficult to show that $X \cup Y$ is a Sperner 1-family, $(X \cup Y) \cup Z$ is a Sperner 2-family, and that every (2,1)-saturated chain partition has $|W|$ chains of length 3 with tops W, and $r(G)$ chains of length 2 with tops some basis of G. Thus the bases of $G_2(P)$ are $W \cup B$, for B a basis of G, so W is a set of isthmuses and $Y \cup Z$ is a set of loops.

REFERENCES

1. Brown, T. J., Transversal theory and F-products, preprint.
2. Crapo, H. and Rota, G.-C., Combinatorial Geometries, MIT Press, Cambridge, Mass. (1970).
3. Dilworth, R. P., A decomposition theorem for partially ordered sets, Ann. of Math. 51 (1950), p. 161-166.
4. Denig, W. A., A class of combinatorial geometries arising from partially ordered sets, Ph.D. dissertation, Ohio State University, (1976).
5. Dowling, T. A. and Kelly, D. G., Elementary strong maps and transversal geometries, Discrete Mathematics 7 (1974), 209-224.
6. Fulkerson, D. R., A note on Dilworth's theorem for partially ordered sets, Proc. Amer. Math. Soc. 7 (1956), 701-702.
7. Greene, C. and Kleitman, D. J., The structure of Sperner k-families, J. Combinatorial Theory, A 20 (1976) 41-68.
8. Greene, C., Sperner families and partitions of a partially ordered set. Combinatorics: Proceedings of the Advanced Study Institute on Combinatorics held at Nijenrode Castle (ed. by J. H. Van Lint and M. Hall), Amsterdam, Part 2, 91-106.
9. Ingleton, A. W. and Piff, M. J., Gammoids and transversal matroids, J. Combinatorial Theory (B) 15 (1973), 51-68.
10. Mason, J. H., On a class of matroids arising from paths in graphs, Proc. London Math Soc. III, Ser. 25 (1972), 55-74.

TRANSVERSAL MATROIDS AND RELATED STRUCTURES

A.W. Ingleton

Balliol College, Oxford, England

ABSTRACT. There is an extensive body of literature on the subject of transversal matroids (with the partial transversals of a family of sets as independent sets) and the more general theory of matroids induced from graphs. This paper reviews the known results and some still unsolved problems. The topics discussed include: characterizations of transversal matroids and of their presentations, standard operations applied to transversal matroids, gammoids and strict gammoids, base-orderability of matroids. The notion of a complete class of matroids (the gammoids being the simplest such class) is introduced and discussed.

1. Simplicial matroids and transversal matroids

Transversal theory and matroid theory were first interrelated in a paper [35] by Rado in 1942, but it was more than twenty years later when it was first explicitly proved by Edmonds and Fulkerson [18] that the partial transversals of a (finite) family of subsets of a (finite) set form the independent sets of a matroid. The present paper is devoted to a survey of some of the results which have since been discovered about the class $\mathcal{T}$ of such transversal matroids and other closely related kinds of matroid. I shall confine my survey to finite situations, ignoring much interesting work which has been done on infinite generalizations.

I shall not make any attempt to keep to the historical order of development. Instead, as I wish to emphasize the geometrical approach to the theory, I shall begin by describing a geometrical configuration which has no immediately apparent relation to transversal theory. Let

M. Aigner (ed.), Higher Combinatorics, 117-131.

$$V = \{v_1, \ldots, v_r\}$$

be a set of r linearly independent elements of a vector space U and let us think of V as (the set of vertices of) an r-simplex. If K is a k-subset of the index set $I = \{1, \ldots, r\}$ the vertex-subset $V(K) = \{v_i : i \in K\}$ spans a k-dimensional subspace (k-flat) which is a k-face of the simplex ($k = 1, \ldots, r$). A simplex is most naturally visualized in projective or affine space but this view must be taken with caution as, in the sequel, we may need additional elements situated at, but distinct from, the vertices (i.e., situated in 1-faces).

Now let T be a subset of U, disjoint from V, the elements of which are "freely placed" on faces of V. Informally, this means that there are no relations between the elements of

$$E = V \cup T$$

other than those imposed by the dimensions of the faces on which they lie. Formally, an element x of E belongs to the span $\langle X \rangle$ of a subset X of T if and only if either $x \in X$ or there are subsets Y of X and K of I such that

$$\langle Y \rangle = F(K) \text{ and } x \in F(K).$$

The resulting configuration S on E with the independence structure induced by that of U has been called by Brylawski [13] a free-simplicial pregeometry.* I shall drop the "free" and just use the term simplicial matroid (not to be confused with the simplicial geometries of Crapo and Rota [14]); V is the spanning simplex of S.

Such a configuration is easily realized analytically. Let

$$T = \{t_1, \ldots, t_n\}_{\neq}.$$

Each element t_j belongs to a unique smallest face $F(K_j)$ of V. Now assign coordinates $(1,0,\ldots,0)$, $(0,1,0,\ldots,0)$,... over some field Φ to the elements of V and take the i th coordinate of the element t_j of T to be

$$\begin{cases} x_{ij} & \text{if } i \in K_j, \\ 0 & \text{if } i \notin K_j, \end{cases}$$

* Other terms which appear in the literature (all in fact equivalent though in some cases defined quite differently) are: principal matroid (or pregeometry) [7,9], free-simplicial pregeometry with spanning simplex [9], fundamental transversal matroid [3,6], principal transversal pregeometry [9,13], strict deltoid [22]. "Free-simplicial pregeometry" (without the qualification "with spanning simplex") has a different meaning in [9]; "principal pregeometry" has a different meaning in [17].

where $(X_{ij} : j = 1,\ldots,n; i \in K_j)$ is a family of independent indeterminates in a transcendental extension of Φ. If Φ is infinite, or finite and sufficiently large, the X_{ij} can be specialized to take values in Φ without introducing unwanted relations.

The configuration may be represented by a bipartite graph Δ with E as vertex-set and edge-set

$$\{(v_i,t_j) : v_i \in V, t_j \in T, i \in K_j\}.$$

It is then easy to see that the bases of $\underline{S}$ are precisely sets of the form $Y \cup X$ with $Y \subseteq V$, $X \subseteq T$ such that $V \setminus Y$ is linked to X in Δ, i.e., they are the transversals of the family $(B_1,\ldots,B_r)$ of subsets of E, where

$$B_i = A_i \cup \{v_i\},\ A_i = \{t_j \in T : i \in K_j\} \quad (i = 1,\ldots,r).$$

Hence $\underline{S}$ is a transversal matroid - but a special one because of the distinguished basis $V = \{v_1,\ldots,v_r\}$ with the property that $v_i \in B_j$ if and only if $i = j$. Thus simplicial matroids are identical with fundamental (or principal) transversal matroids [3,6,9,13].

However, although $\underline{S}$ may be special as a transversal matroid, the restriction $\underline{M} = \underline{S} \mid T$ is a matroid on T with independent sets the partial transversals of the family $(A_1,\ldots,A_r)$ and this is clearly a completely arbitrary family. Hence we have the result (first explicitly stated in this form by Brylawski [13]):

THEOREM 1. A matroid is transversal if and only if it can be obtained from a simplicial matroid by deleting the spanning simplex.

This theorem is also inherent in a result of the author's [19,20] but the geometrical description was there phrased in rather different, and less illuminating, terms. One can even say that it was essentially proved (though in analytic rather than geometrical terms) by Mirsky and Perfect who in 1967 established the linear representability of transversal matroids by assigning indeterminate coordinates exactly as described above [29 - and see 28, Theorem 7.1.3].

2. Characterizations of transversal matroids

To study the finer structure of transversal matroids we need to look at the cyclic flats. In a given matroid $\underline{M}$ a set X is called cyclic, or fully dependent, if it is a union of circuits, that is, if its complement is a flat of the dual matroid $\underline{M}^*$. In particular, X is a cyclic flat of $\underline{M}$ if and only if the complement of X is a cyclic flat of $\underline{M}^*$. It is not difficult to check that a matroid $\underline{M}$ on a given set is completely determined by specifying the collection $\underline{F}(\underline{M})$ of all its cyclic flats together with their ranks (see, e.g., [13]). This is often a very economical way of describing a particular matroid. $\underline{F}(\underline{M})$ has a natural lattice structure, with $F \vee G$ the usual join of flats, i.e. the span of $F \cup G$, and $F \wedge G$ the union of all the circuits contained in $F \cap G$. A pleasant feature of this lattice is that one only has to invert it to obtain $\underline{F}(\underline{M}^*)$.

It is implicit in the definition of a simplicial matroid $\underline{S}$ that a cyclic set of rank k in $\underline{S}$ or in the restriction $\underline{S}|T$ must lie in a k-face of V; moreover, a cyclic k-flat must be a k-face (or, more precisely, the intersection of that face with E or T as the case may be). This fact not only shows that the number of cyclic k-flats in a transversal matroid of rank r cannot exceed $\binom{r}{k}$ as was proved by Brualdi and Mason [10] but also imposes a restriction on the way the cyclic flats fit together.

THEOREM 2. A matroid $\underline{M}$ of rank r is transversal if and only if there is an injective map

$$\phi : \underline{F}(\underline{M}) \longrightarrow \underline{2}^r$$

(the lattice of subsets of $\{1,\ldots,r\}$) such that

(1) $\quad |\phi F| = \rho F \qquad (\forall F \in \underline{F}(\underline{M}))$,

(2) $\quad \phi(F \vee G) = \phi F \cup \phi G \qquad (\forall F,G \in \underline{F}(\underline{M}))$,

(3) $\quad \rho \bigcap F_j \leq |\bigcap \phi F_j| \qquad$ for every finite family of cylic flats F_j.

[ρ denotes the rank function of the matroid $\underline{M}$.]

A criterion of this kind was first given by Mason [24], but in terms of the collection of all cyclic sets rather than just cyclic flats.

It is important to notice that, to within permutations of $\{1,\ldots,r\}$, there can be at most one ϕ satisfying (1) and (2). This makes it relatively straightforward to apply Theorem 2 directly to test whether a given matroid is transversal (cf. the algorithm in [8]). However, Mason also deduced from it an entirely intrinsic characterization of transversal matroids in terms of inequalities to be satisfied by ρ. If $(F_1,\ldots,F_N)$ is a family of cyclic flats and $J \subseteq \{1,\ldots,N\}$, let $F(J)$ denote $\bigcup(F_j : j \in J)$.

THEOREM 3. A given matroid is transversal if and only if, for every family $(F_1,\ldots,F_N)$ of cyclic flats,

$$\rho(F_1 \cap \ldots \cap F_N) \leq \sum_{J \subseteq \{1,\ldots,N\}} (-1)^{|J|+1} \rho F(J).$$

Mason's original theorem [24] stated the condition in terms

of all families of cyclic sets, but, as in the case of Theorem 2, it is not difficult to show that it is enough to consider cyclic flats*. In fact, it is possible to go further and require the condition only for a very restricted set of families of cyclic flats.

DEFINITION. Call a set A a critical flat if it is the intersection of all the cyclic flats which properly contain it; then let $F_1,\dots,F_N$ be the minimal (in terms of set inclusion) members of the set of all cyclic flats properly containing A and put

$$\delta A = \sum_{J\subseteq\{1,\dots,N\}} (-1)^{|J|+1}\rho F(J) - \rho A .$$

[It is easy to see that δA is also given by the same sum taken over the family of all cyclic flats properly containing A.]

We then have

THEOREM 4. A given matroid is transversal if and only if $\delta A \geqslant 0$ for every critical flat A.

Another analytic criterion for a transversal matroid was first obtained quite independently [22] via duality from the theory of strict gammoids which I shall be discussing later in §5.

DEFINITION. The set function β in a matroid $\underline{M}$ of rank r is defined recursively by the formula

$$\beta X = r - \rho X - \sum \beta F$$

where the sum is taken over all cyclic flats F properly containing X.

THEOREM 5. $\underline{M}$ is transversal if and only if $\beta X \geqslant 0$ for every set X.

This turns out, perhaps not surprisingly, to be closely related to Theorems 3 and 4. In fact it is clearly sufficient if $\beta A \geqslant 0$ for every critical flat A, and with the aid of Möbius inversion in the lattice $\underline{F}(\underline{M})$ it is not hard to prove the

PROPOSITION. For every critical flat A, $\beta A = \delta A$.

3. Presentations

A transversal matroid can usually be "presented" by more than

* Indeed, if one really wants to achieve maximum economy at some expense of clarity this whole discussion can be confined just to those cyclic flats whose complements are indecomposable flats of $\underline{M}^*$ in the sense of [22, §4] - see Theorem 8 below.

one family of subsets. It is well known (and has been tacitly assumed in §2) that presentations $(A_1,\ldots,A_r)$ of length r equal to the rank of the matroid always exist, and I shall confine this discussion to such presentations. (Longer presentations can occur only when the matroid has coloops.) The presentation is called minimal if replacing any A_i by a proper subset leads to a different matroid, maximal presentations are defined similarly.

The simplicial embedding described in §1 is helpful in characterizing the possible presentations of a transversal matroid $\underline{M}$ on T. If T intersects each (r-1)-face of the simplex V in a hyperplane (of $\underline{M}$) then the presentation $(A_1,\ldots,A_r)$ corresponding to the simplicial embedding is clearly minimal and the A_i are cocircuits (the complements in T of the faces). It was shown conversely by Bondy and Welsh [3] that in any minimal presentation the A_i are cocircuits. Thus it is always possible to find a simplicial embedding such that the (r-1)-faces actually correspond to hyperplanes of $\underline{M}$. The resulting configuration of hyperplanes was called a quasi-simplex in [19,20].

In general there are several minimal presentations but there is, apart from order, only one maximal presentation [2,8,22]. If (A_i) is any presentation, let C_i be the span of A_i in $\underline{M}^*$, i.e. $T\setminus C_i$ is the largest cyclic flat contained in $T\setminus A_i$. If we use (C_i) to define a simplicial embedding by reversing the procedure of §1 it is clear that the resulting matroid has the same cyclic flats and is therefore the same matroid. Thus (C_i) is still a presentation and is evidently maximal. Uninqueness follows from the uniqueness of the mapping ϕ of Theorem 2: C_i is just $T\setminus F_i$ where F_i is the largest cyclic flat such that $i \notin \phi F_i$.

These observations are the basis of the algorithm given by Brualdi and Dinolt [8] which tests whether a given matroid is transversal and produces the maximal presentation if it is.

Finally, we note an obvious consequence of this discussion:

THEOREM 6. A transversal matroid of rank r has a unique presentation if and only if it has r cyclic hyperplanes.

4. Subclasses of $\mathcal{T}$

We have already encountered the subclass $\mathcal{S}$ of simplicial matroids. If $\underline{S}$ is simplicial then $\underline{S}^*$ is also simplicial (dualization just corresponds to interchanging the roles of V and T in the bipartite graph Δ [22]) and therefore certainly transversal. Thus, to use an obvious symbolism,

$$\mathcal{S} \subset \mathcal{T} \cap \mathcal{T}^*.$$

This inclusion is certainly proper - for instance all transversal matroids of rang ≤ 3 are also cotransversal [22]. There seems to be no way of characterizing the class $\mathcal{T} \cap \mathcal{T}^*$ of transversal-cotransversal matroids except by analytic criteria derived from Theorems 3-5.

In terms of presentations, we may define a hierarchy of subclasses

$$\mathcal{T}_1 \subset \mathcal{T}_2 \subset \mathcal{T}_3 \subset \dots$$

of $\mathcal{T}$. A transversal matroid will be assigned to $\mathcal{T}_k$ if there is a (minimal) presentation $(A_1,\dots,A_r)$ such that the intersection of every k+1 of the A_i is empty. This is equivalent to saying that every element lies on a k-face in the corresponding simplicial embedding.

The matroids of $\mathcal{T}_1$ with presentations by a disjoint family are of little interest. $\mathcal{T}_2$ has recently been identified rather unexpectedly by Matthews [26] with the class of bicircular matroids of Simoes-Pereira [37]. These are matroids defined on the edge-sets of graphs (with loops and multiple edges), the circuits of the matroid being "bicycles" - edge-sets of subgraphs homeomorphic from one of

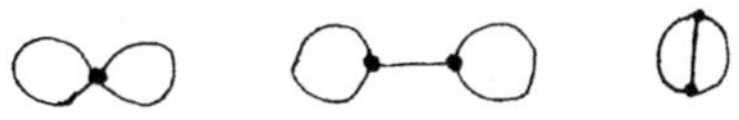

It is natural to ask whether there is any analogous interpretation of $\mathcal{T}_k (k \geqslant 3)$ in terms of hypergraphs.

Related to, and including, $\mathcal{T}_1$ is the class of <u>partitional matroids</u> studied by Recski [36] in which distinct <u>sets</u> A_i in the presentation are disjoint but may be repeated in the family $(A_1,\dots,A_r)$.

It is known that all transversal matroids are representable by linear dependence over suitably large fields (see §1 above and [33]). One might seek to characterize those matroids of $\mathcal{T}$ which are representable over some particular field. For instance, binary transversal matroids were shown by de Sousa and Welsh [16] to coincide with graphic transversal matroids which had already been characterized by Bondy [1]. I am not aware of any other results in this area.

5. <u>Duality - strict gammoids</u>

Let us now return to the situation described in §1 of a transversal matroid $\underline{M} = \underline{S}|T$ obtained as the restriction of a simplicial matroid S with spanning simplex $V = \{v_1,\dots,v_r\}$. We may suppose, as remarked in §3, that r is the rank of $\underline{M}$ and we may then label the elements of T as $t_1,\dots,t_n$ so that $t_i \in A_i$ $(i=1,\dots,r)$. Now, from the bipartite graph Δ associated with $\underline{S}$ we construct a digraph Γ with vertex-set T by drawing an arc from t_i to t_j in Γ whenever v_i is joined to t_j in Δ. We then see that an r-subset X of T is linked to V in Δ if and only if $T \setminus X$ is linked in Γ (by vertex-disjoint paths, possibly of zero length) to $B_o = \{t_{r+1},\dots,t_n\}$ [22, Lemma 3.1]. Hence the subsets of T linked in Γ to B_o are the bases of a matroid $\underline{M}(\Gamma, B_o)$ which is just the dual $\underline{M}^*$ of $\underline{M}$.

Matroids obtained in this way from digraphs were called by Mason [25] strict gammoids. The construction is readily reversed to show that every strict gammoid is obtainable as the dual of a transversal matroid. So we have

THEOREM 7 [22,27]. The class $\mathcal{T}^*$ of cotransversal matroids is identical with the class of strict gammoids.

Dualization of matroids has no nice geometrical interpretation in general, but we can obtain an equivalent interpretation of strict gammoids which has some intuitive geometrical content as follows. If $\underline{M} = \underline{S}|T$, then the dual $\underline{M}^* = \underline{M}(\Gamma, B_o)$ is the contraction $\underline{S}^* \cdot T$. Now, as observed in §4, $\underline{S}^*$ is also a simplicial matroid with spanning simplex T and V a set of r independent points freely placed on the faces of T. Thus $\underline{M}(\Gamma, B_o)$ may be visualized as the result of projecting an n-simplex T in an n-dimensional vector space U from the r-flat R spanned by an independent set V freely placed in faces of V (or more formally as the image of T in the quotient-space U/R). This description is equivalent to a theorem of Dowling and Kelly [17, Theorem 3.3] which identifies what they call a "principal pregeometry" with a cotransversal matroid.

Any characterization of the matroids of class $\mathcal{T}$ (e.g. Theorems 2 - 5) translates by duality into a characterization of the matroid of class $\mathcal{T}^*$. In particular, we should perhaps state explicitly the dual-analogue of Theorem 5 which was in fact first obtained by Mason [25] in the context of strict gammoids. To state this in the form which involves the least calculation, we need some more matroid terminology from [22]. A flat F is called decomposable if $F = F_1 \cup F_2$ where F_1, F_2 are proper subflats of F such that

$$\rho F_1 + \rho F_2 = \rho F \quad \text{and} \quad \rho(F_1 \cap F_2) = 0;$$

F is called indecomposable otherwise. (Thus, in the loopless case, "indecomposable" is equivalent to "connected".) An indecomposable flat is certainly cyclic, but not conversely .

DEFINITION. In a given matroid $\underline{M}$, define a set-function α recursively by the formula

$$\alpha X = |X| - \rho X - \sum \alpha F$$

where the sum is taken over all indecomposable flats F properly contained in X.

THEOREM 8 [25,22]. A matroid is cotransversal if and only if $\alpha A \geqslant 0$ for every set A which is a non-trivial union of indecomposable flats.

Essentially the same result, but in terms of cyclic flats is given by Brualdi [7,(4.10)].

6. Gammoids

Let A, B_0 be subsets of the vertex-set T of a digraph Γ. It was proved by Perfect [31] and in greater generality by Pym [34] that the subsets of A which are linked in Γ to subsets of B_0 are the independent sets of a matroid on A, which we call a gammoid. Clearly such a gammoid is just the restriction $\underline{M}(\Gamma,B_0)|A$ of a strict gammoid. Hence the fact that gammoids are matroids follows from the discussion in §5. Also, since strict gammoids are just contractions of simplicial matroids, we have

THEOREM 9 [22]. The class $\mathcal{G}$ of gammoids is precisely the class of minors of simplicial matroids, and hence is closed under the operations of restriction, contraction and dualization.

The relations between the main classes of matroids which we have been discussing are summarized in the following diagram.

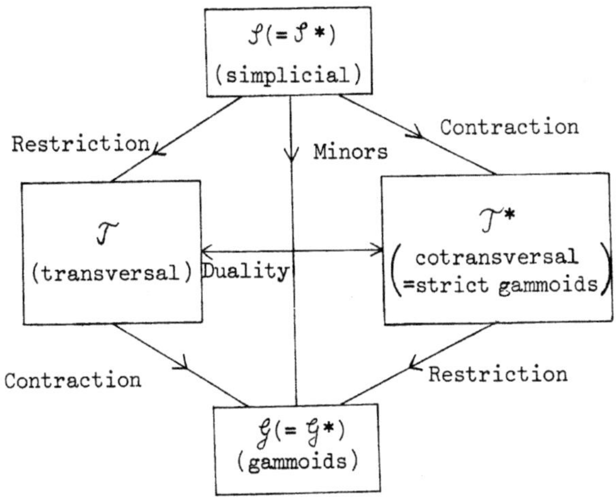

A gammoid is a particular case of a more general construction. If $\underline{M}_0$ is a matroid on the vertex-set T of a digraph Γ (or on some subset of T) and if $A \subseteq T$, then the subsets of A linked in Γ to independent sets in $\underline{M}_0$ are the independent sets of a new matroid $\underline{M}_1$ on A [5,25,32]. Let us call this process of obtaining $\underline{M}_1$ from $\underline{M}_0$ induction. A simple graphical construction shows that, as an operation on classes of matroids, induction is a closure operation. But, by definition, $\mathcal{G}$ is the class obtained from free matroids by induction and hence is itself closed under induction. Now, if a gammoid $\underline{M}$ is induced from the free matroid on B_0, the truncation to rank k of $\underline{M}$ is induced from the k-uniform matroid on B_0. But uniform matroids are obviously gammoids (in fact transversal). So we have:

THEOREM 10. The class $\mathcal{G}$ is closed under induction and truncation.

Let us call a class of matroids complete if it is closed under the operations of restriction, contraction, dualization, direct sum, truncation and induction. (The same concept could be defined by various different lists of permissible operations, depending on personal taste. This list certainly includes some logically redundant items.)

The only one of these operations not already mentioned in relation to gammoids is that of direct sum. But it is obvious that a direct sum of gammoids is still a gammoid. We may therefore state finally:

THEOREM 11. $\mathcal{G}$ is a complete class of matroids.

In fact, $\mathcal{G}$ occupies a distinguished position in the hierarchy of complete classes. A moment's thought shows that it is the smallest non-empty complete class. For any such class certainly contains all matroids of rank 0 (by induction using a graph with no edges); hence, by duality, all free matroids; and hence, by induction, all gammoids.

It is an important outstanding problem to find an intrinsic characterization of gammoids in general. There are some partial results covering very special cases:

THEOREM 12 [22]. A matroid of rank at most 3 is a gammoid if and only if it is cotransversal.

[In fact, all matroids of rank ≤ 2 are cotransversal.]

THEOREM 13 [12,13,20]. For a binary matroid $\underline{M}$, the following assertions are equivalent.

(a) $\underline{M}$ is a gammoid.

(b) $\underline{M}$ has no minor isomorphic to $\underline{M}(K_4)$-the cycle matroid of the complete graph K_4.

(c) $\underline{M}$ is isomorphic to the cycle matroid of a series-parallel network.

Since binary matroids may be characterized as those which do not have as a minor the 4-point line, Theorem 13 provides an "excluded minor" criterion for binary gammoids. There are a number of famous "excluded minor" theorems in matroid theory and perhaps the next to look for in the present context is one for ternary gammoids (i.e. gammoids representable over GF(3)). I believe that it is probably futile to look for a nicely determined complete list of excluded minors to characterize $\mathcal{G}$ as a whole. Such a list would certainly have to be infinite; there are even infinitely many minimal non-gammoids of rank 3 (easily recognized in view of Theorems 8 and 12) and no apparent means of cataloguing all those. It may well be that, although relatively simple "excluded minor" criteria have been found for classes of matroids which are closed under restriction and contraction but not much more, one should seek exclusions in a different sense to characterize complete classes. I shall return to that topic in §8.

7. Subclasses of $\mathcal{G}$

I have used the term "induction" in the wide sense, covering digraphs in general. In the special situation where $\underline{M}_1$ is induced from $\underline{M}_0$ via a bipartite graph, let us say that $\underline{M}_1$ is derived from $\underline{M}_0$. Thus, if $\mathfrak{F}$ is the class of free matroids, then $\mathcal{T}$ may be identified with $\mathfrak{F}'$, the class of all derivations of free matroids. Repeating the process we obtain a hierarchy of classes by successive derivation:

$$\mathfrak{F} \subset \mathfrak{F}' \subset \mathfrak{F}'' \subset \dots \ .$$

The matroids in the class

$$\mathfrak{F}^{(\infty)} = \bigcup_{r=1}^{\infty} \mathfrak{F}^{(r)}$$

were called cascades by Mason [25] who gave an example of a (strict) gammoid which is not a cascade. It would be interesting to find characterizations of the classes $\mathfrak{F}^{(r)}$ $(r \geqslant 2)$ and/or $\mathfrak{F}^{(\infty)}$.

Another elusive subclass consists of those gammoids obtainable from undirected graphs. As far as induction of matroids is concerned, an undirected graph is equivalent to the digraph obtained by replacing each undirected edge by a pair of arcs in opposite directions. Woodall has given an example of a gammoid which is not obtainable from any undirected graph in a paper [38] in which he also discusses other variations and generalizations of the concept of graphic induction. It is interesting that this example happens to be both a strict gammoid and a cascade.

The class $\mathcal{G}$ is closed under truncation (and so is $\mathcal{T}^*$), but $\mathcal{S}$ and $\mathcal{T}$ are not. Matroids which are truncations of simplicial matroids have been characterized by Brualdi and Dinolt [9]. As far as I am am aware, there is no known characterization of the truncations of $\mathcal{T}$. These certainly form a proper subclass of $\mathcal{G}$; it is easily checked that the rank-3 (strict) gammoid

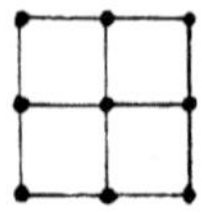

(which also happens to be Mason's example of a non-cascade) is not a truncation of any transversal matroid.

8. Base-oderability and other complete classes

We have observed that $\mathcal{G}$ is the smallest non-trivial complete class. Two larger complete classes are determined by concepts of base-orderability which were introduced by Brualdi [4] in relation to transversal matroids.

A matroid $\underline{M}$ is strongly base-orderable if, given any two bases B_1, B_2, there is a bijection $\phi : B_1 \longrightarrow B_2$ such that, for every subset X of B_1,

(*) $(B_1 \setminus X) \cup \phi X$ is also a base.

$\underline{M}$ is called just base-orderable if the exchange property (*) can be satisfied at least for subsets X of cardinal 1 and r-1 (where r is the rank of $\underline{M}$).

A series of results [4,5,11,15,16,22,25] have shown, in effect, that the classes $\mathcal{BO}$ of bare-orderable, and $\mathcal{SBO}$ of strongly base-orderable, matroids are complete. So we have inclusions

$$\mathcal{G} \subset \mathcal{SBO} \subset \mathcal{BO}.$$

It is well known that the first inclusion is strict; the simplest example to demonstrate the fact is

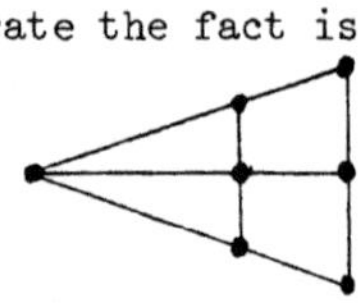

which is a non-gammoid by Theorems 8 and 12, but strongly base-orderable by Theorem 14 below. That the second inclusion is also strict is shown by an example in [21]. In the same paper it is shown that the excluded-minor problem has no tidy solution for the class $\mathcal{BO}$.

As with $\mathcal{G}$, it is only in the two special cases of rank-3 and binary matroids that a complete picture of the classes $\mathcal{BO}$ and $\mathcal{SBO}$ exists. All matroids of rank ≤ 2 are strongly base-orderable; in rank 3, base-orderable and strongly base-orderable are equivalent by definition and $\underline{M}(K_4)$ is the only obstruction (see, e.g., [15,21]):

THERQREM 14. A matroid of rank 3 is (strongly) base-orderable if and only if it has no restriction isomorphic to $\underline{M}(K_4)$.

In the binary case, $\underline{M}(K_4)$ is again the only obstruction:

THEOREM 15 [16]. A binary matroid is base-orderable if and only if it has no minor isomorphic to $\underline{M}(K_4)$.

So, combining Theorems 13 and 15 we see that, for binary matroids, being base-orderable, strongly base-orderable and a gammoid are all equivalent.

There is scope for introducing an infinity of complete classes between $\mathcal{BO}$ and $\mathcal{SBO}$ by appropriate limitations on the cardinals of subsets X for which (*) is to hold, but these do not seem to have been studied.

Other familiar complete classes are the class $\mathcal{R}$ of all representable matroids and subclasses corresponding to representability over fields of specified characteristic. Las Vergnas has also remarked recently that the orientable matroids [23] form a complete class.

As an alternative to seeking complete classes defined by some common property of the member matroids, one can define complete classes by sets of generators, or "non-generators". Given a family of matroids $(\underline{M}_1,\ldots,\underline{M}_k)$, let $\mathcal{C}(\underline{M}_1,\ldots,\underline{M}_k)$ denote the smallest complete class of matroids which contains all the $\underline{M}_i$; let $\mathcal{E}(\underline{M}_1,\ldots,\underline{M}_k)$ denote the largest complete class which contains none of the $\underline{M}_i$.

Since the available evidence suggests that minimal families of excluded minors for interesting complete classes are always infinite, perhaps the obstruction question for complete classes should be asked in the following form.

PROBLEM 1. For a given complete class $\mathcal{C}$ of matroids (in particular for $\mathcal{C} = \mathcal{G},\mathcal{BO},\mathcal{SBO}$) can one find a finite family $(\underline{M}_1,\ldots,\underline{M}_k)$ such that

$$\mathcal{C} = \mathcal{E}(\underline{M}_1,\ldots,M_k)?$$

Or, instead of obstructions, one can look for generators:

PROBLEM 2. For a given complete class $\mathcal{C}$ can one find a finite family $(\underline{M}_1,\ldots,\underline{M}_k)$ such that

$$\mathcal{C} = \mathcal{C}(\underline{M}_1,\ldots,\underline{M}_k)?$$

This question has a trivial answer for $\mathcal{G}$ as we have seen (the unique matroid on the empty set will serve as a generator), but appears to be quite open in the other cases.

The special negative significance which attaches in this theory to $\underline{M}(K_4)$ - the simplest matroid which is not a gammoid -

suggests the following problem from the reverse point of view.

PROBLEM 3. Identify the class $\mathcal{E}(\underline{M}(K_4))$.

One possible answer to Problem 3 would be $\mathcal{BO}$, making $\underline{M}(K_4)$ the unique obstruction in the present sense to base-orderability. At the opposite extreme, it may be that the class $\mathcal{K}$ of matroids which do not have $\underline{M}(K_4)$ as a minor is already complete, in which case the answer to Problem 3 would be $\mathcal{K}$. Both these answers may well be false; but they cannot both be true, since it is shown in [21] that $\mathcal{K}$ is strictly larger than $\mathcal{BO}$. The first answer would be the most satifying but I conjecture that the second is the correct one.

REFERENCES

1. J.A. Bondy, Transversal matroids, base-orderable matroids and graphs, Quart.J.Math.(Oxford)(2) 23 (1972), 81-89.
2. ———, Presentations of transversal matroids, J.London Math.Soc.(2) 5 (1972), 289-92.
3. ——— and D.J.A. Welsh, Some results on transversal matroids and constructions of identically self-dual matroids, Quart.J.Math.(Oxford)(2) 22 (1971), 435-51.
4. R.A.Brualdi, Comments on bases in dependence structures, Bull.Australian Math.Soc. 7 (1969), 161-67.
5. ———, Induced matroids, Proc.American Math.Soc. 29 (1971), 221-31.
6. ———, Fundamental transversal matroids, Proc.American Math. Soc. 45 (1974) 151-56.
7. ———, Matroids induced by directed graphs, a survey, Proc. 2nd Czech.Graph Theory Symp., Prague (1974),115-34.
8. ——— and G.W. Dinolt, Characterizations of transversal matroids and their presentations, J.Comb.Theory 12 (1972), 268-86.
9. ——— ———, Truncations of principal geometries, Discrete Math. 12 (1975), 113-38.
10. ——— and J.H. Mason, Transversal matroids and Hall's theorem, Pacific J.Math. 41 (1972), 601-13.
11. ——— and E.B. Scrimger, Exchange systems, matchings and transversals, J.Comb.Theory 5 (1968), 244-57.
12. T.H. Brylawski, A combinatorial model for series-parallel networks, Trans.American Math.Soc. 154 (1971), 1-22.
13. ———, An affine representation for transversal geometries, Studies in Appl.Math. 54 (1975),143-60.
14. H.H. Crapo and G.-C. Rota, Combinatorial Geometries (M.I.T. Press, 1970).
15. J. Davies, Some problems in matroid theory, D.Phil.thesis, Univ. of Oxford (1975).

16. J. de Sousa and D.J.A. Welsh, A characterization of binary transversal structures, J.Math.Anal.Appl. 40 (1972), 55-59.
17. T.A. Dowling and D.G. Kelly, Elementary strong maps and transversal geometries, Discrete Math. 7 (1974), 209-24.
18. J.R. Edmonds and D.R. Fulkerson, Transversals and matroid partition, J.Res.Nat.Bur.Standards 69B (1965), 147-53.
19. A.W. Ingleton, Conditions for representability and transversality of matroids, in Théorie des Matroides, ed.C.P.Bruter (Springer, 1971), 62-66.
20. ————, A geometrical characterization of transversal independence structures, Bull.London Math.Soc. 3 (1971), 47-51.
21. ————, Non-base-orderable matroids, Proc. 5th British Combinatorial Conf., Aberdeen (1975), 355-60.
22. ———— and M.J. Piff, Gammoids and transversal matroids, J.Comb.Theory 15 (1973), 51-68.
23. M. Las Vergnas, Matroides orientables, preprint.
24. J.H. Mason, A characterization of transversal independence spaces, in Théorie des Matroides, ed.C.P.Bruter (Springer, 1971), 86-94.
25. ————, On a class of matroids arising from paths in graphs, Proc.London Math.Soc. (3) 25 (1972), 55-74.
26. L.R. Matthews, Bicircular matroids, preprint.
27. C. McDiarmid, Strict gammoids and rank functions, Bull.London Math.Soc. 4 (1972), 196-98.
28. L. Mirsky, Transversal Theory (Academic Press, 1971).
29. ———— and H. Perfect, Applications of the notion of linear independence to problems in combinatorial analysis, J.Comb. Theory 2 (1967), 327-57.
30. H. Narayanan and M.N. Vartak, Gammoids, base-orderable matroids and series-parallel networks, preprint.
31. H. Perfect, Applications of Menger's graph theorem, J.Math. Anal.Appl. 22 (1968), 96-110.
32. ————, Independence spaces and combinatorial problems, Proc.London Math.Soc. (3) 19 (1969), 17-30.
33. M.J. Piff and D.J.A. Welsh, On the vector representation of matroids, J.London Math.Soc. (2) 2 (1970), 284-88.
34. J.S. Pym, The linking of sets in graphs, J.London Math.Soc. 44 (1969), 542-50.
35. R. Rado, A theorem on independence relations, Quart.J.Math. (Oxford) 13 (1942), 83-89.
36. A. Recski, On partitional matroids with applications, Proc. Colloq.Math.Soc.János Bolyai, Keszthely (1973), 1169-79.
37. J.M.S. Simões-Pereira, On subgraphs as matroid cells, Math.Z. 127 (1972), 315-22.
38. D.R. Woodall, Linking in graphoids, preprint.

MATROIDS AS THE STUDY OF GEOMETRICAL CONFIGURATIONS

J. H. Mason

Faculty of Mathematics, Open University, U.K.

ABSTRACT. Matroids arise in a variety of combinatorial and algebraic contexts. The ideas of

- independence and bases as in vector spaces
- dependence as in algebraic dependence
- circuits as in graphs
- flats as in projective geometries
- atomic semimodular lattices,

all come down to the same underlying structure. The essential feature of matroids is that they extract from a situation the basic incidence properties of points, lines and planes which we associate with geometry. That they show up in so many places and disguises suggests them as worthwhile objects of study. How that study is carried out is affected to some extent by the direction of approach. Graph theory suggests certain ideas to be generalised, lattice theory suggests others, and vector spaces still others. Often it seems that the geometrical aspect is lost in a profusion of algebraic notation.

My intention therefore is to review the basic matroid constructions from a geometric point of view. This will act as an introduction for those not familiar with matroids, and I hope it will afford some insight for experts not used to thinking geometrically. In order to preserve the geometric-intuitive nature of the lectures, I have included very few explicit theorems, and fewer proofs. Nevertheless, in most cases there is enough information to verify the assertions without recourse to references.

M. Aigner (ed.), Higher Combinatorics, 133-176.

1. MATROIDS, CONSTRUCTIONS AND MAPS

Categorically, matroids are a little strange in that there are two kinds of maps (strong and weak) which are studied. Neither category provides a rich enough setting for describing matroid constructions.

1.1 Objects - Matroids
1.2 Basic Constructions - Restrictions and Extensions
  - Contractions and Lifts
1.3 Inducing Matroids from Paths in Graphs
  - Transversal Matroids
  - Matroid Joins
1.4 Maps - Strong
  - Weak
1.5 Strong and Weak Maps Together
  - Higgs Factoring
  - Truncating and Erecting
1.6 A Geometric Picture of Duality

2. DILWORTH TRUNCATION AND GENERALIZATIONS

This standard geometrical construction seems to have been little studied, yet it has the potential of a powerful tool for matroid theory.

2.1 The Dilworth Truncation
2.2 Generalizations
2.3 Relation to Matroid Joins
2.4 Linear Representability
2.5 Lattice Approach
2.6 Generalized Coordinates.

1. MATROIDS: CONSTRUCTIONS AND MAPS

1.1 Objects - Matroids

We are interested in geometrical configurations like the following:

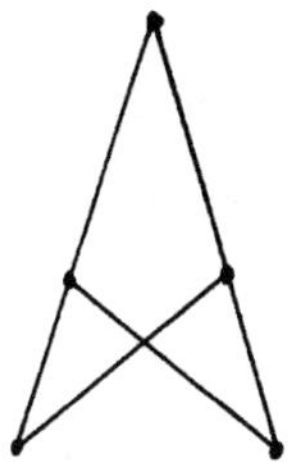

K_4

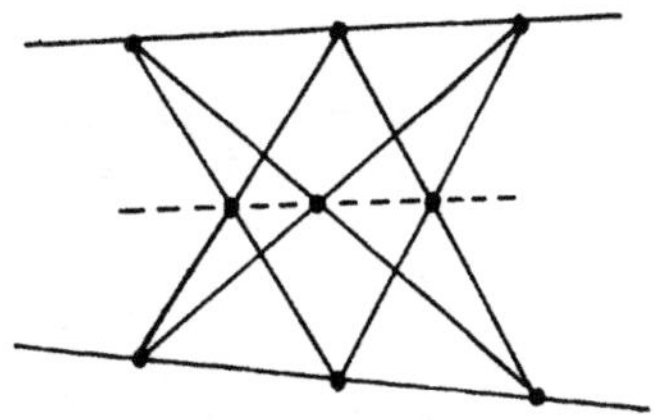

PAPPUS (with dotted line)
PAPPUS$^-$ (without)

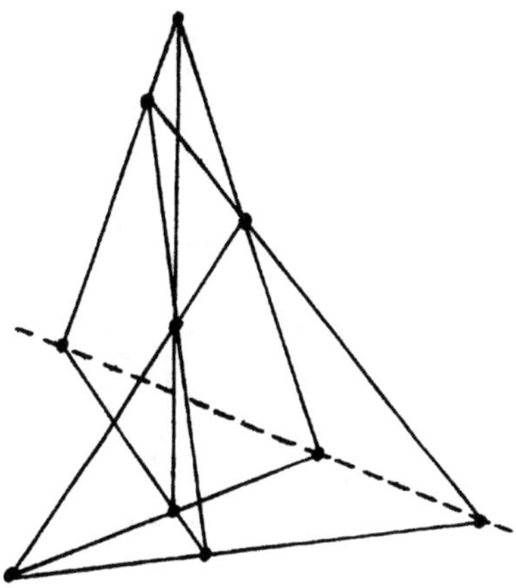

DESARGUES
DESARGUES$^-$ (without dotted line)

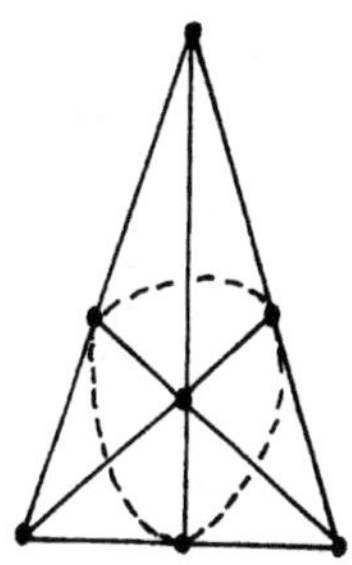

FANO
FANO$^-$

Figure 1 Planar Configurations - only lines with more than two points are shown explicity.

They consist of a (finite) set P of points, and some subsets of P called lines, planes and so on whose incidence properties are geometrical. By this I mean that:

Any 2 points lie on a unique line
Any 3 points (not colinear) lie on a unique plane
(or perhaps more to the point - any line and point not on it lie on a unique plane)

And so on. Notice that there is no assumption that lines in a plane must meet.

In traditional geometry, points have affine dimension one. We wish to encompass both this traditional approach and the vector space situation in which a one-dimensional subspace has many points (vectors). Therefore I shall begin by speaking of a collection of flats with associated rank, each consisting of a set of points. We define a Combinatorial Geometry as a finite set of flats with associated rank k, $0 \leq k$ such that:

any k-flat and 1-flat not in it, lie in a unique k+1 flat.

The following examples are useful to get to know because they illustrate some of the things that can happen, and are good examples for testing out theories. They are based on a matroid V_4 called the Vamos configuration since it was used by Vamos in questions concerning linear representability.

	3-Flats with more than 3 points	4-Flats with more than 4 points	5-Flats with more than 5 points
V_4 (Vamos-rank 4)	1234 1256 3456 5678 3478	12345678	
V_4^+ rank 4	Same as V_4 with 1278	12345678	
V_5 rank 5	Same as V_4^+	123456 123478 125678 345678	12345678

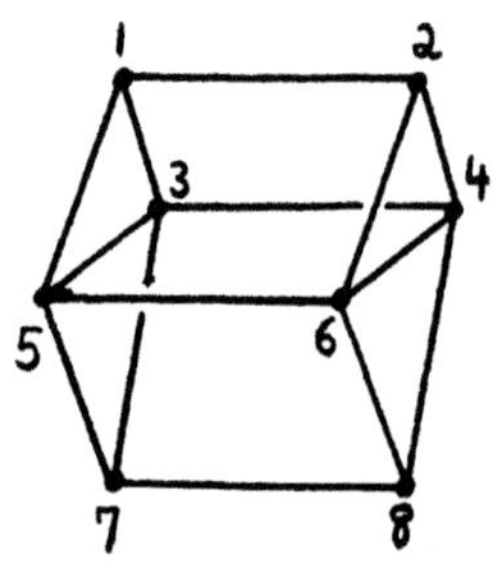

Figure 2 The Vamos configuration

Because of the limitations on sketching high dimensional figures in the plane, I use the same sketch to represent each of these three matroids.

There are many different ways to describe incidence, depending on the principal source of your examples. The major ones are listed below.

1935	H.Whitney [19]	Matroid	generalizing linear independence especially motivated by forests in graphs represented in a vector space over GF(2).
1936	S. Maclane [10]	Geometric Lattice	Lattice of flats of projective, affine and other geometries
1937	B.L.Van der Waerden [17]	Dependence	Linear and algebraic dependence

Each approach to incidence involves a set of axioms concerning some abstracted concept such as basis, dimension, dependence, independence and so on, and for finite sets, they all turn out to be equivalent. We can use whichever concepts are most handy in a given situation - bewildering for the uninitiated but fun for the expert. Details can be found in Crapo-Rota [3], Brylawski [1] and Welsh [18].

The following table summarizes the major concepts which we shall need.

MATROID CONCEPTS	LINEAR	GRAPHS	NOTATION	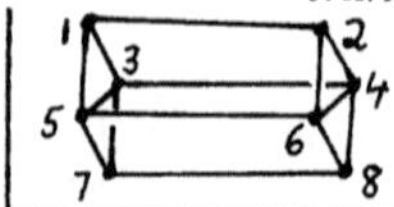
points S	vectors	edges	S,T,U,V	{1,2,3,4,5,6,7,8}
Independent sets $\mathcal{I}$. $\phi\in\mathcal{I}$, $A\subseteq B\in\mathcal{I}\Rightarrow A\in\mathcal{I}$. $\|A\|<\|B\|$ and $A,B\in\mathcal{I}\Rightarrow$ $\exists\, b\in B\backslash A$. . $A\cup\{b\}\in\mathcal{I}$ Dependent= not independent	Linearly independent Steinitz Exchange	Forest	A,B indep	Samples {1} {1,2} {1,2,3} {1,2,3,5} {1,2,7,8}
Base ≡ Maximal Independent set	Basis	Spanning Forest	B	{1,2,3,5} {1,2,7,8}
Circuits $e\in C_1\cap C_2$, $C_1\neq C_2$ $\exists\, C\subseteq C_1\cup C_2\setminus\{e\}$	Minimal dependent	circuit	C,D	{1,2,3,4} {5,6,7,8} {1,2,5,6}
Rank ρ integer valued $\rho(\phi)=0$, $\rho(\{x\})\leq 1$, $A\subseteq B\Rightarrow\rho A\leq\rho B$, $\rho A\cup B+\rho A\cap B\leq$ $\rho A+\rho B$	Dimension	size of spanning forest	ρ,rk	$\rho V_4=4$ $\rho(\{1,2,3,4\})=3$ $\rho(\{1,2\})=2$ $\rho(\{1\})=1$
Flats Maximal sets with given rank. Ordered by inclusion= Geometric lattice: atomic, semi-modular	Subspaces (Affine subspaces)		F,G,H	{1,2,3,4} {3,4,5,6} {7,8}
Hyperplanes= flats of rank $\rho S-1$	Hyperplanes	Complements of cocircuits		{1,2,3,4}
Closure of A ≡smallest flat containing A	Span		<A>	<{1,2,5}> ={1,2,5,6}

Lower case letters denote elements of sets and $S\cup e$ means $S\cup\{e\}$.

The name Combinatorial Geometry is used when we work with the geometric lattice of flats. However when we are considering sets of points then we find that often the 1-flats are sets of points, and hence we have a pre-geometry or matroid. The transition from matroid to lattice mirrors the transition from a vector space to projective or affine geometry. The situation is particularly graphic in graphs. In graph theory it is sometimes useful to permit multiple edges and loops. In the corresponding cycle matroid, the loops make up the 0-flat, and the parallel edges between two fixed vertices (together with all loops) form a 1-flat.

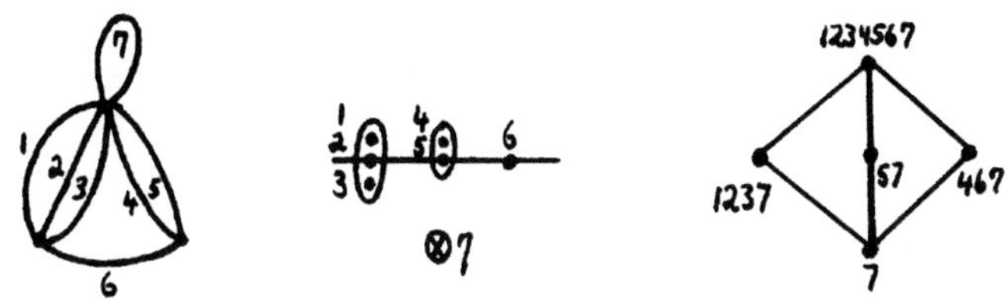

Notation: ⊗ denotes a loop

(:) denotes parallel points

Figure 3 Loops and parallel edges

Since many matroid constructions introduce parallel points and loops, it is best to permit them from the start.

Examples:

1. Free matroids. A matroid on S which has no circuits is called free, and denoted by F(S), or F_n when only the cardinality is important.

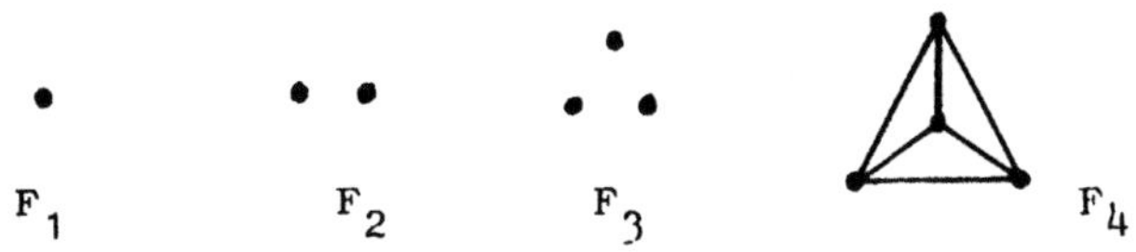

Figure 4 Free matroids

More generally we use the word free to mean "as independent as possible". For example we speak of a point e lying freely on a flat F.

2 and 3 are free on $\langle 1,2,3\rangle$ but 1 is not.
6 is free on $\langle 1,2,3,4,5,6\rangle$

We also speak of M(S) being <u>free-er</u> than N(S), meaning that every subset A of S independent in N is also independent in M.

2. Uniform matroids. The uniform matroid of rank k on n points- denoted by $U_k(n)$ has as bases all k-sets and as circuits all k+1-sets.

Figure 5 Uniform matroids

<u>Linear Representability</u>

Since a chief motivating example for matroids was subsets of vector spaces it is natural to ask when a matroid arises in this way. We say that a matroid M(S), without loops, is <u>representable</u> in V over the field K is there is a one-one function

$$f: S \longrightarrow V$$

such that

$$A \text{ indep in } M \Leftrightarrow f(A) \text{ linearly indep in } V.$$

The Desargues configuration of rank 3 is clearly not representable by well known theorems of projective geometry. The Desargues configuration in rank 4 cannot have a "broken line" for the same reason as in vector spaces - the intersection of two planes is at most a line.

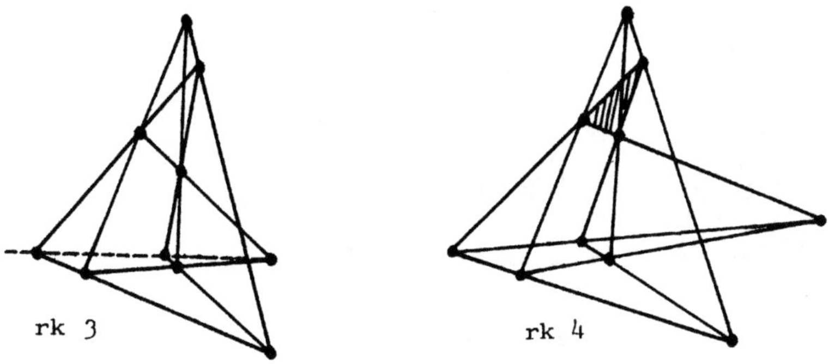

Figure 6 Desargues in rank 3 and Desargues in rank 4.

However we can easily arrange a non-representable matroid in rank 4 - namely the Vamos configuration V_4, on eight points (the smallest possible). In any vector space there would be a point lying on all four lines <1,2>, <3,4>, <5,6>, <7,8> so that <1,2> and <7,8> would have to be coplanar. In fact V_4 and Desargues are closely related as we shall see shortly.

Before going further it will be useful to have a picture of the general directions of matroid theory as they have developed recently.

1. Distinguishing and Characterizing classes of examples e.g. Graphic, Binary, Transversal, Regular, Orientable...
2. Constructions of new matroids from old.
3. Counting e.g. Chromatic, Tutte, Whitney polynomials.
4. Applications e.g. Matching theory and Network Analysis.

In these two lectures I shall concentrate on 2.

1.2 Basic Constructions

<u>Restriction</u>: By restricting our attention to a subconfiguration of M, we obtain a submatroid. This works because we have not required flats to intersect modularly, as in projective geometry, and it justifies calling matroid theory the study of geometrical configurations.

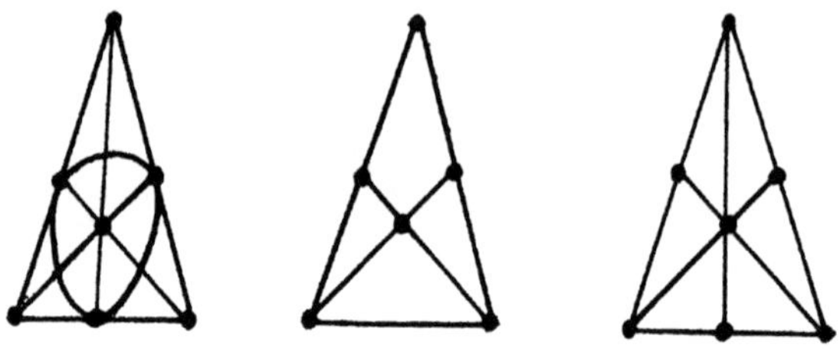

Figure 7 Fano, a restriction and a non-restriction

If M is a matroid on S and V is a subset of S then the <u>restriction</u> of M to V, denoted by M|V is specified by:

A ≤ V indep in M|V ⟺ A indep in M
rk A is unchanged

It is also convenient to write M\V to denote the restriction of M to the complement of V, in other words, M delete V.

<u>Contraction:</u> The name comes from graphs-contracting an edge. In vector spaces it corresponds to quotients, and the geometrical equivalent is projection.

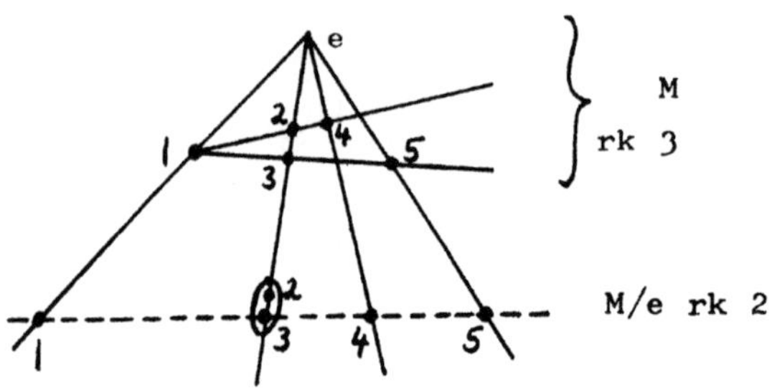

Figure 8 Contraction of M by e seen as projection

If M is a matroid on S and V is a subset of S then put U = S\V. We specify the contraction of M by V as a matroid on U denoted by M/V:

A indep in M/V $\Leftrightarrow$ A $\cup$ B indep in M for some (any) base B of M|V.

rk A = $\rho(A\cup V) - \rho V$

Contraction by V is geometrically thought of as a sequence of projections from points in some basis of V.

Theorem: The operations of restriction and contraction commute

$$(M/A)\,|\,(B\backslash A) = (M|\,(B\cup A))/A$$

M contracted by A, restricted to B\A = M restricted to B $\cup$ A contracted by A.

This is an easy exercise in rank functions, more difficult using bases or circuits, and even more difficult in lattices where it is virtually the same as the Scum Theorem (Crapo-Rota [3]).

Exercise $(M|\,(B\cup A))/A \neq (M|B)/(A\cap B)$

Note: It is traditional when given a matroid M(S) and a point e in S, to think of M and M/e as two entirely distinct matroids. Yet the geometric idea of projection suggests that M and M/e are parts of one larger matroid. The formal construction of this larger matroid is a special case of the Dilworth truncation which I will discuss in the second lecture.

Dual: The name comes from graphs. Whitney [19] showed that the idea of a planar dual could be generalized to matroids (and hence to arbitrary graphs), and he provided a geometric interpretation when the matroids arise from a vector space. By the end of this lecture we will be in a position to generalize this to arbitrary matroids.

Let M be a matroid on S. Then the dual of M, denoted M*(S) is specified by:

A $\leq$ S is indep in M* $\Leftrightarrow$ S\A contains a base of M
(Thus bases of M and M* are complementary)
$\rho^*A = |A| - [\rho(S) - \rho(S\backslash A)]$

size of base of M

size of least intrusion of a base of M into A

Largest base-complement of M meeting A

Hyperplanes of M* are complements of circuits of M.

Examples: $U_k^*(n) = U_{n-k}(n)$

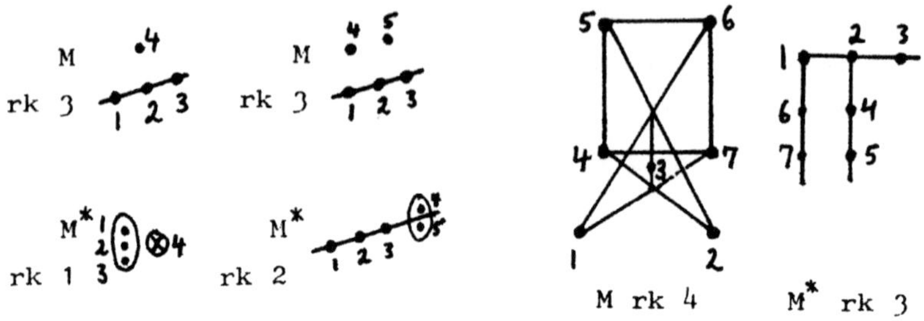

Figure 9 Three matroids and their duals.

It is routine to check that $M^{**} = M$, though it is not true if you look only at lattices of flats because often M^* will have parallel elements, so the lattice loses information.

Theorem (Tutte [16]) The dual of contraction is restriction

$$(M/V)^* = M^* \backslash V$$

This is a straightforward exercise in the manipulation of rank functions, circuits, or bases, as you prefer!

The operations of restriction and contraction can be reversed. The description of how to do this represents a major contribution by Crapo and Higgs respectively.

Restriction-Extension: If instead of deleting points we wish to adjoin them, then we must consider what information we need to specify. Geometrically we wish to know which old lines, planes and flats generally are to contain the new point.

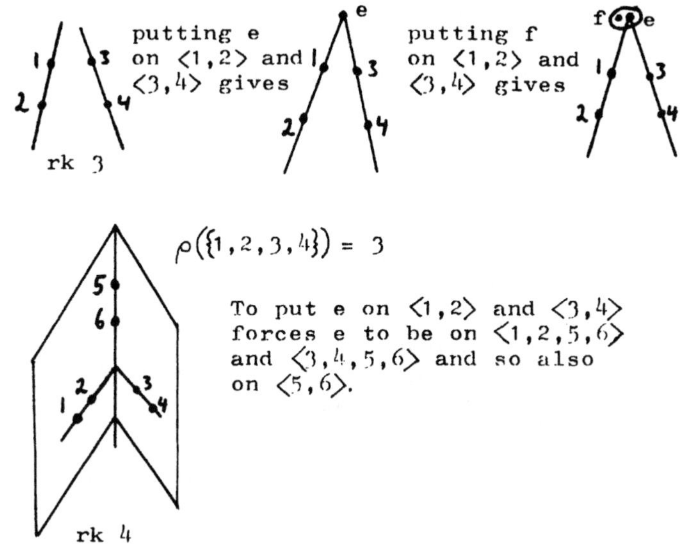

Figure 10 Examples of extensions

A <u>modular cut</u> $\mathcal{M}$ for M(S) is a set of flats such that

(i) $F \le G$ and $F \in \mathcal{M} \Rightarrow G \in \mathcal{M}$ (Filter)
(ii) $F, G \in \mathcal{M}$ and $\rho F + \rho G = \rho(F \cup G) + \rho(F \cap G)$
$\Rightarrow F \cap G \in \mathcal{M}$ (Modular pairs)

The necessity of the modular pairs condition (ii) is illustrated in the diagram above.

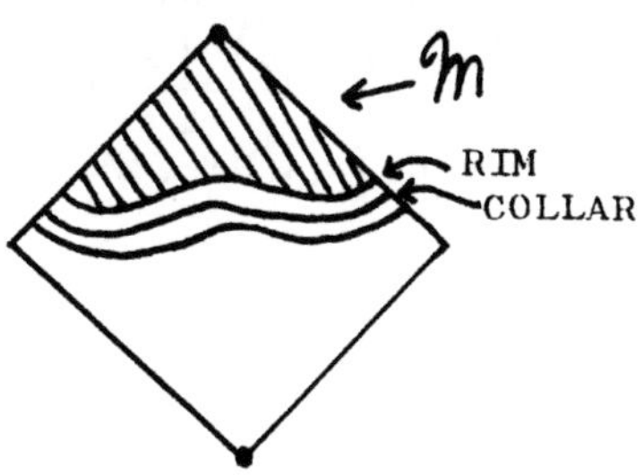

Each flat in the collar is covered by a unique member of $\mathcal{M}$

Figure 11 A typical modular cut

Given a modular cut $\mathcal{M}$ we can specify an extension of M denoted $M(S \cup e;\mathcal{M})$ or M^e for short, with rank function ρ^e, by

$$A \cup e \text{ is indep in } M^e \Leftrightarrow A \text{ is indep in } M \text{ and } \langle A\rangle \notin \mathcal{M}$$

$$A \le S \text{ is indep in } M^e \Leftrightarrow A \text{ is indep in } M$$

$$\rho^e(A) = A \quad A \le S$$

$$\rho^e(A \cup e) = \begin{cases} \rho A \text{ if } \langle A\rangle \in \mathcal{M} \\ \rho A+1 \text{ if } \langle A\rangle \notin \mathcal{M} \end{cases}$$

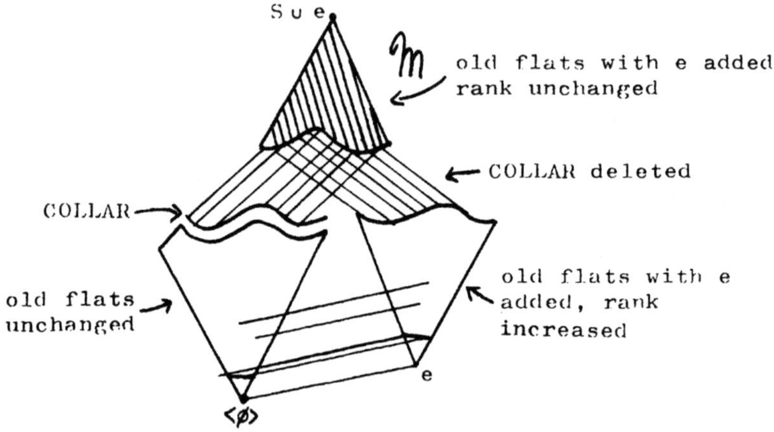

Figure 12 One point extension

Examples

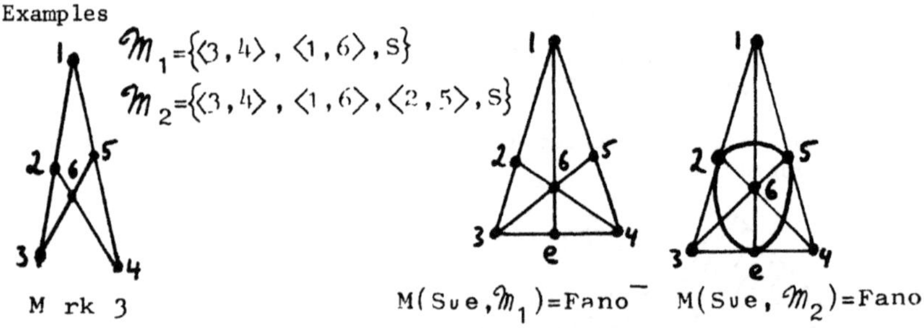

Figure 13 Fano and Fano⁻ as extensions of K_4.

Special Cases

(i) By choosing the empty modular cut we have the free extension M + e. Here e belongs to every basis, and the total rank has increased by one.
(ii) By choosing a principal filter (all flats containing a given flat F), we obtain a modular cut which corresponds geometrically to placing e as freely as possible on the old flat F.

The one point extensions of M(S) (we usually exclude M + e and consider only the extensions with unchanged rank), form a lattice, since the set of modular cuts of M is closed under intersections. The zero-element is formed by adding a loop by means of the principal modular cut generated by <ϕ>. The one-element is the addition of a point freely in M(S) (without increasing the rank) by the principal modular cut with one element, S.

Theorem Contractions and extensions commute.
Given two modular cuts $\mathcal{M}_1$ and $\mathcal{M}_2$ for M(S) we can extend by $\mathcal{M}_1$, then form the modular cut $\mathcal{M}_2^+$ generated by $\mathcal{M}_2$ in the extension. (Take the intersection of all modular cuts of M^{e_1} containing the flats of $\mathcal{M}_2$.) As we shall see shortly, the order in which the extensions are made usually makes a difference.

Contraction - Lift: To reverse the operation of contraction we need, given M(S), to construct N(S∪e) such that N/e = M. Duality provides one answer:

$$\begin{aligned} & N/e = M \\ \text{iff}\quad & (N/e)^* = M^* \\ \text{iff}\quad & (N^*\backslash e) = M^* \\ \text{iff}\quad & N^* \text{ is a one point extension of } M^*. \end{aligned}$$

This solves the problem in principle, but much more can be said once we have the notion of matroid join, so I shall postpone the geometrical discussion.

1.3 Inducing Matroids From Paths in Graphs

This method of constructing matroids lies at the heart of matching problems and transversals. It is usually treated quite independently of restriction-extension and contraction-lift, yet it is just another way of looking at some special cases. Nevertheless, I find that inducing from a graph, particularly a bipartite graph, provides a useful iconic representation of the standard constructions, especially for geometries of large rank.

Therefore it is worth spending enough time on it to get a firm grasp of the ideas.

The primary idea is this: Given a bipartite graph Γ on $V \times S$, and a matroid $M(S)$, then a new matroid $M^{\Gamma}(V)$ is induced on V by:

A is indep in $M^{\Gamma}(V)$ $\Leftrightarrow$ A is linked in Γ
(by pairwise vertex disjoint paths)
to an indep set in M.

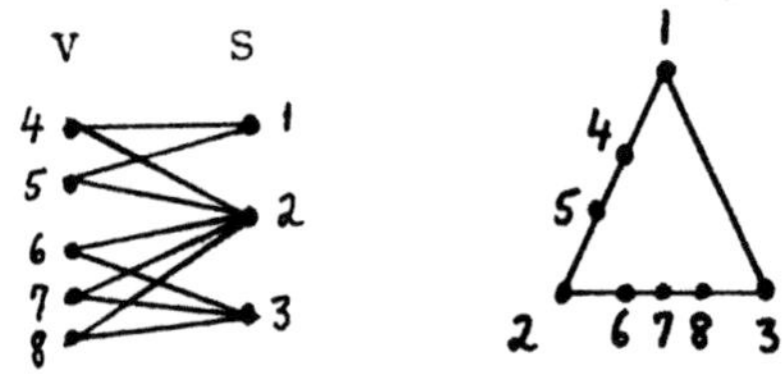

Figure 14 A matroid induced from a bipartite graph

The first studies had M(S) the free matroid on S. Independent sets in M^{Γ} are then the partial transversals of the family of subsets $(\{v : (v,s) \in \Gamma\};\ s \in S)$ of S (Edmonds and Mirsky). Then Rado observed that any matroid could be placed on S. Several people (Perfect, Pym, Brualdi) showed that Γ could be replaced by an arbitrary directed graph.

An important observation which assists both the analysis and the geometric picturing is that not only do we get M^{Γ} induced on V, but we actually have a matroid on $V \cup S$ whose restriction to S is M(S) and whose restriction to V is $M^{\Gamma}(V)$. This encompassing matroid is specified by

$A \leq V \cup S$ is indep $\Leftrightarrow$ A is linked in Γ
to an indep set in M(S).

Ingleton [8] used this idea to describe transversal matroids geometrically. We begin by displaying S as a simplex in $|S| - 1$ dimensional Euclidean space - a representation of the free matroid on S. Now place the elements of V as points on the faces of the simplex, the point v being placed freely on the face

$$\Gamma(V) = \{s : (v,s) \in \Gamma\}$$

of the simplex. The result is the matroid on $V \cup S$, and the restriction to V is the induced matroid M^{Γ}. (This is illustrated in the previous diagram.)

The same idea carries over more generally for the matroid induced on $S \cup e$ by Γ is the principal extension of M(S) by the modular cut generated by the flat $\Gamma(e)$.

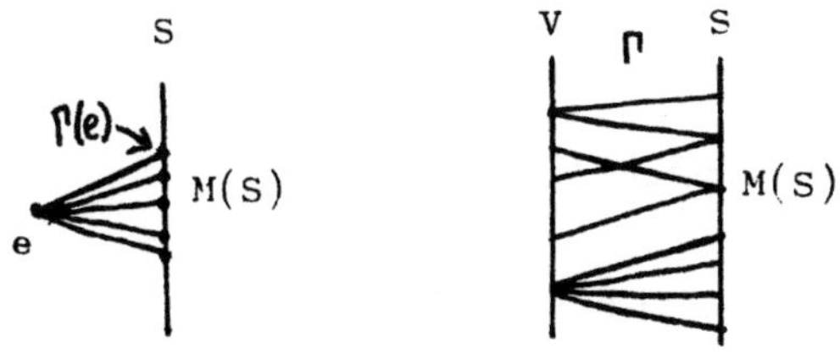

Figure 15 Principal extensions.

Thus the matroid induced on $V \cup S$ is the result of a number of principal extensions made simultaneously. We can deduce that any number of principal extensions can be performed simultaneously and the result is independent of the order in which they are carried out.

This raises an interesting question. Is there an analogue in linking of bipartite graphs for non-principal extensions? Of course there is (it concerns the use of several graphs Γ_i on $V \times S$ and is a messy restatement of modular cuts), but is there a neat graphical statement? The question is bound up with the following technical, but important question.

> Given M(S) and modular cuts $\mathcal{M}_1, \ldots, \mathcal{M}_t$ for M, under what conditions are these modular cuts compatible in the sense that there exists $N(S \cup E)$ such that for each i, $1 \leq i \leq t$,
>
> $$N|(S \cup e_i) = M(S) \text{ extended by } \mathcal{M}_i.$$

A neatly phrased answer to this question could have application in a number of aspects of matroid theory. So far I have partial results only.[1] As an example of the difficulties that arise, consider the following:

[1]This has been answered by Hien Nguyen in his M.I.T. Thesis 1975.

$$\mathcal{M}_1 = \{\langle 1,2\rangle, \langle 3,4\rangle, S\}$$
$$\mathcal{M}_2 = \{\langle 1,2\rangle, \langle 3,4\rangle, \langle 5,6\rangle, S\}$$

Figure 16 Incompatible modular cuts

$\mathcal{M}_1$ and $\mathcal{M}_2$ are not compatible because $\mathcal{M}_1$ followed by $\mathcal{M}_2$ forces e_2 to be a loop. Doing $\mathcal{M}_2$ followed by $\mathcal{M}_1$ forces e_1 to lie on $\langle 5,6\rangle$ as well.

Applications of Inducing

1. We can now examine the lift construction geometrically. Recall that $N(S \cup e)$ is a lift of M if $N/e = M$. The free-est lift of M is formed by adjoining a new point e to form $M + e$ and then lifting each point s along the line $\langle s,e\rangle$. (See below.) This is precisely the matroid you get by inducing. First take a copy S_1 of S. Then induce as shown.

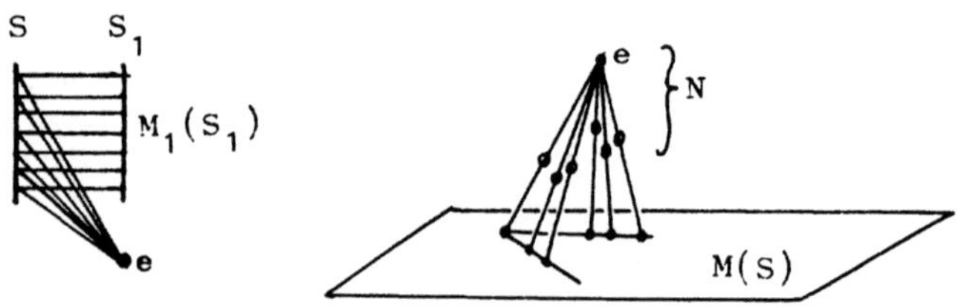

Figure 17 The lift

The matroid N on $S \cup e$ is the matroid we are after.

<u>Theorem</u> The matroid $N(S \cup e)$ is the free-est matroid on $S \cup e$ whose contraction by e is M.

Proof: Suppose $Q(S \cup e)/e = M$. Let $A \leq S$ be indep in $Q(S \cup e)$. Then either:

(i) A is indep in M, in which case A is indep in N by construction.

(ii) A is dependent in M, in which case there exists a circuit $C \leq A \cup e$ in Q. By a circuit exchange argument C is the unique such circuit. Thus C\e is the unique circuit in M contained in A. But then A is indep in N by construction. (Link one element of C\e to e, the rest of $M(S_1)$)

This result illustrates the fact that if $M^{\Gamma}(V \cup S)$ is induced by Γ and M(S), and if $A \leq S$, then the induced matroid contracted by A, $M^{\Gamma}(V \cup S)/A$ is the same as the matroid induced by Γ and M(S)/A. (To be indep in the contraction by A, a set must be extendable by some basis of A to an indep set.)

2. This extremely important application is due independently to Nash Williams [13] and Edmonds [6]. Despite its being frequently used, it has not been all that well studied, and it poses almost as many problems as it solves. The easiest way to begin is with a well known special case.

Matroid Sum: Given M(S) and N(T) with $S \cap T = \phi$, we can specify a new matroid M + N on $S \cup T$ by

$$A \text{ is indep in } M + N \Leftrightarrow \begin{array}{l} A \cap S \text{ is indep in } M \text{ and} \\ A \cap T \text{ is indep in } N. \end{array}$$

$$\rho A = \rho_M(A \cap S) + \rho_N(A \cap T).$$

Geometrically M and N are completely free of each other, which is reminiscent of the direct product of vectorspaces. However M + N is categorically a co-product not a product. The free matroid F(S) is the sum of free matroids on each singleton of S.

Matroid Joins: Now we bring in the bipartite graph. Given $M_1(S)$ and $M_2(S)$ we replace S by two distinct copies S_1 and S_2 with the corresponding matroids $M_1(S_1)$ and $M_2(S_2)$. Now induce the join $M_1 \vee M_2(S)$ from the graph Γ as shown:

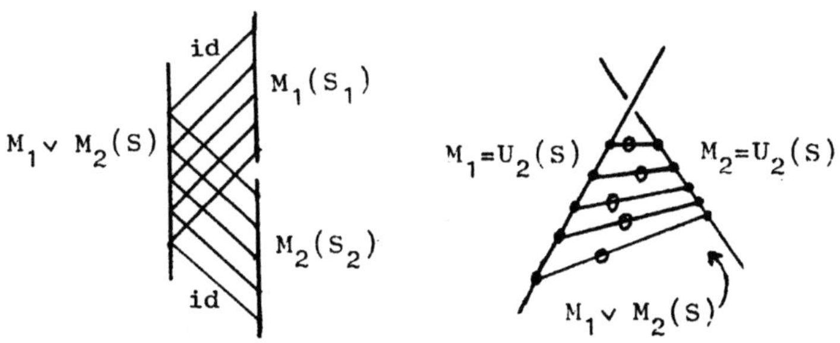

Figure 18 Matroid Joins

Here Γ restricted to $S \times S_i$ is the identity $S \to S_i$. The new matroid $M_1 \vee M_2$ has:

$$A \text{ indep in } M_1 \vee M_2 \Leftrightarrow A = A_1 \cup A_2 \text{ with } A_i \text{ indep in } M_i(S)$$

$$\rho A = \underset{B \leq A}{\text{Max}} [\rho_1 B + \rho_2(A \backslash B)]$$

The geometrical interpretation is that we form $M_1 + M_2$, then for each $s \in S$ we put a new point freely on the line $<s_1, s_2>$. It is often useful to think of $M_1 \vee M_2$ as being on the set $S \cup S_1 \cup S_2$, because then it is easy to see that this larger matroid has the property that

$$\big((M_1 \vee M_2)/S_1\big)|S = M_2(S)$$

$$\big((M_1 \vee M_2)/S_2\big)|S = M_1(S).$$

As an example, the lift construction that we looked at earlier using bipartite graphs can be shown as a matroid join, by noticing that $U_1(S)$ is effectively the same matroid (same lattice of flats) as $F(e)$, so that $M \vee U_1 = N$.

A very natural and rather more symmetrical matroid construction which includes matroid join, is the following. Given $M_1(S)$ and $M_2(T)$ we specify a new matroid $M_1 \nabla M_2$ on $S \times T$ by inducing it from $M_1(S) + M_2(T)$ using the graph Γ whose vertices are $(S \times T) \cup S \cup T$. The vertex (s,t) is joined in Γ to s in S and t in T.

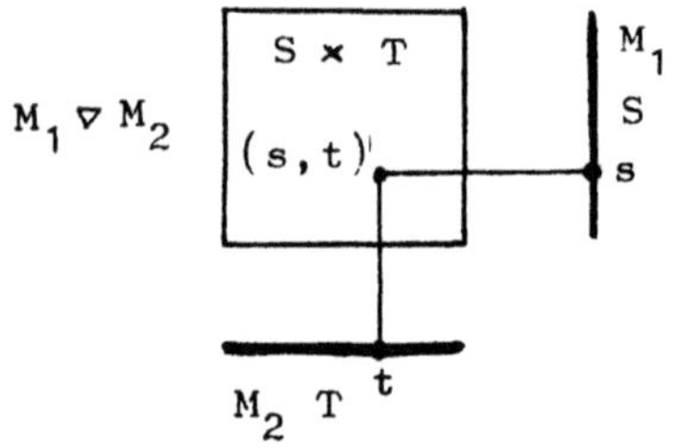

Figure 19 $M_1 \nabla M_2$ on $S \times T$

Then

$$A \text{ is indep in } M_1 \nabla M_2 \Leftrightarrow A \text{ is linked in } \Gamma \text{ to an indep set in } M_1 + M_2.$$

I suspect that $M_1 \nabla M_2$ might have quite interesting properties.

1.4 Maps

The question arises as to what maps should look like in a category of geometrical configurations. For suggestions and inspiration we can look at

(i) constructions we have already seen:

extensions = embeddings
contractions = quotients

(ii) analogies from vector spaces, graphs, lattices and so on
(iii) something that will provide a rich structure.

There are currently two accepted categories of matroids, with rather uncategorical properties. I shall outline the situation here, and in the second lecture indicate some steps towards an alternative.

Strong Maps

Coming from the direction of (ii) in the lattice context, Higgs and Crapo reached the idea of a strong map:

$\sigma : S \to U$ is a strong map $M(S) \to N(U)$

$\Leftrightarrow$ $\sigma^{-1}(F)$ is a flat of M for all flats F of N.

Examples

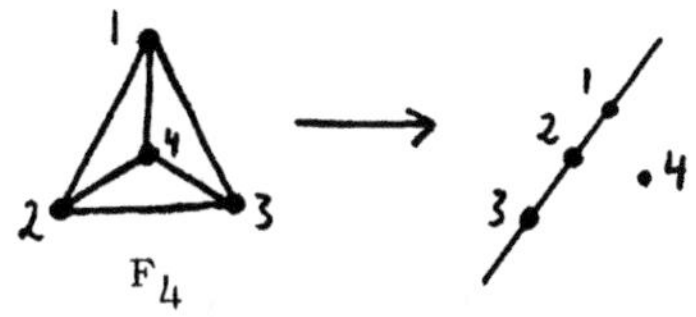

1. Figure 20. The identity $S \to S$ is a strong map $F(S) \to M(S)$.

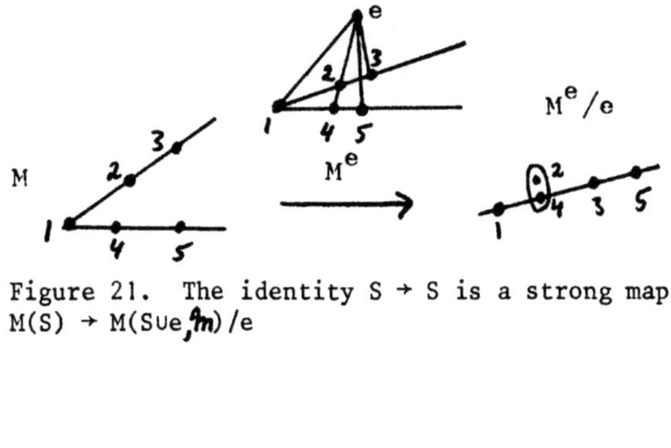

2. Figure 21. The identity $S \to S$ is a strong map $M(S) \to M(S \cup e, m)/e$

Special cases include truncation when $\mathcal{M} = \{S\}$. Here a point e is added freely to the flat S in M, and then we project from e. The truncation of M is denoted by M^T.

3. The identity $S \to S$ is a strong map $M(S) \to M/V + U_0(V)$ What is happening here is that we need somewhere to send the elements of $V \leq S$, so we adjoin them back onto M/V as loops, described by the uniform rank zero matroid on V. There is a strong case to be made for specifying M/V as a matroid on S not just on $S \setminus V$ since it corresponds more closely to vector space quotients. There the kernel becomes the 0-dimensional space. The quotient space is really defined on the same set as the domain, but subspaces are now cosets.

Although these examples seem special, in that both domain and codomain are matroids on the same set S, we can always arrange that this will happen, and that the function is the identity, by the following devices. If $f : S \to T$ then replace both S and T by the set $S \cup (T\setminus f(S))$. Then a matroid M(S) is replaced by $M(S) + F(T\setminus f(S))$ and N(T) is replaced by a matroid with the same lattice of flats as N, but with the point t replaced by the parallel points $f^{-1}(t)$.

Weak Maps

Higgs introduced the idea of "free-er than" in the form of a map (called weak to distinguish it from strong). The identity map $S \to S$ is a weak map $M(S) \to N(S)$ if

$$\text{A indep in N} \Rightarrow \text{A indep in M}$$
$$\rho_M(A) \geq \rho_N(A)$$

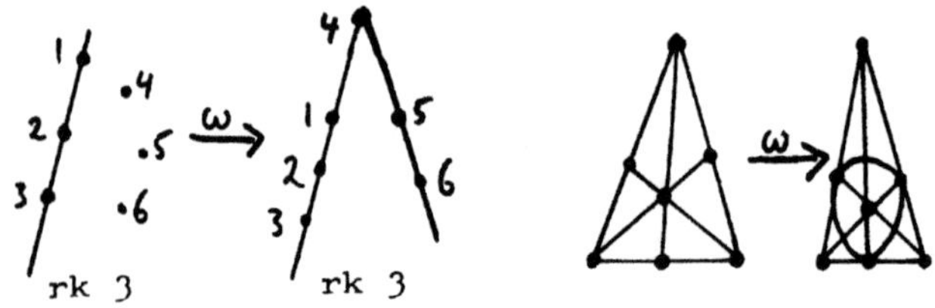

Figure 22 Examples of weak maps

Notice that every strong map is weak - this is most easily seen by observing that

$$\sigma \text{ is a strong map} \Leftrightarrow \text{ for any sets } X \geq Y,\ \rho_M X - \rho_M Y \geq \rho_N X - \rho_N Y$$

whereas

$$\sigma \text{ is a weak map} \Leftrightarrow \text{ for any set } X,\ \rho_M X - \rho_M \phi \geq \rho_N X - \rho_N \phi.$$

In future σ and τ will denote strong maps and ω a weak (not necessarily strong) map.

Exercise: Any weak map is a truncation followed by a rank preserving weak map.

It seems to be very difficult to describe the structure of weak maps. Despite repeated attempts there is no real progress towards making precise the geometrical intuition that weak maps seem to be locally (on subsets) like strong maps. Again, the free matroid and matroid sums are the only categorical constructions which work.

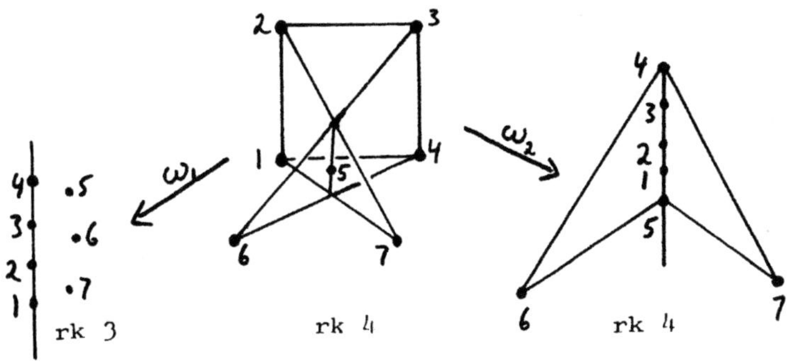

Figure 23 Showing that there need not be a free-est weak image that decreases the rank of a set A = {1,2,3,4} by one

1.5 Strong and Weak Maps Together

On their own, strong maps and weak maps do not form rich enough categories to permit us to use categorical ideas. Yet working together, there are some results of a canonical nature. Often it is the case that a diagram of strong maps describes a matroid construction which is either the free-est or least-free such object. In other words objects in the category of strong maps tend to be canonical with respect to weak maps. One reason for this is that rank preserving strong maps are isomorphisms, yet frequently there are several matroids of the same rank which satisfy a given strong map diagram, and which are related weak maps. I shall illustrate this situation with some examples.

Higgs Factoring [7]

Theorem Given a strong map $\sigma : M \to N$ then σ can be factored as a sequence of strong maps each of degree 1. Further, the factors are free-er than the factors of any other factoring.

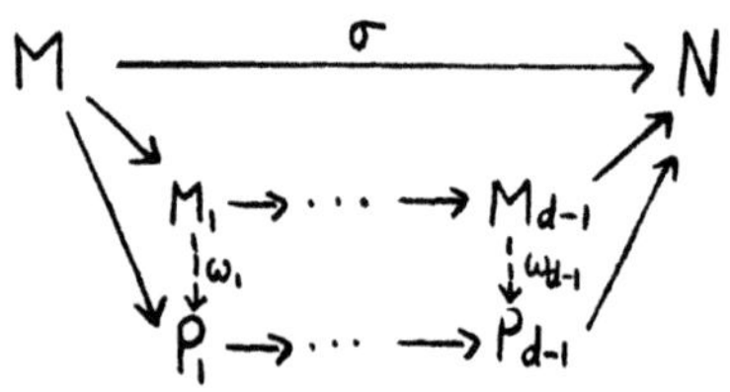

rk M − rk N = d

Figure 24 Higgs Factoring

Proof Outline: The most direct approach is to observe that

$$\{F : \rho_M F - \rho_N F = \text{rk } M - \text{rk } N\}$$

forms a modular cut for M. Making the extension and then contracting produces the M_1 of the theorem. At each stage M_{i+1} is constructed from M_i in the same way. The fact that contracting and extending commute gives the triangular diagram.

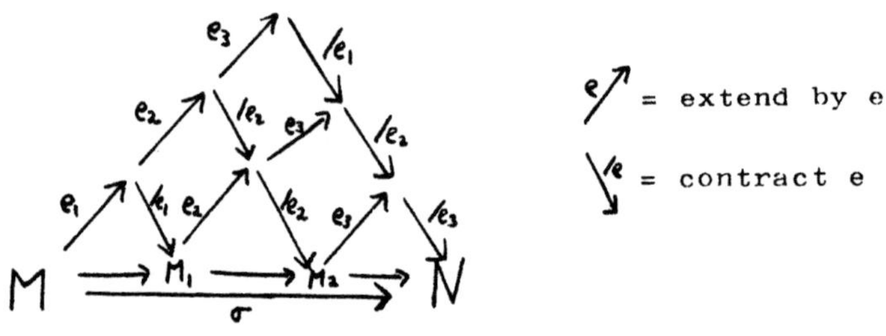

Figure 25 The Mapping Triangle

The canonical part is deduced from the following lemma.

If rk P = rk Q and $\begin{array}{ccc} P(S) & \xrightarrow{\sigma} & N(S) \\ {\scriptstyle\omega}\downarrow & \nearrow{\scriptstyle\tau} & \\ Q(S) & & \end{array}$ then $\begin{array}{ccccc} P & \to & P_1 & \to & N \\ {\scriptstyle\omega}\downarrow & & \downarrow{\scriptstyle\omega_1} & \nearrow & \\ Q & \to & Q_1 & & \end{array}$

where P_1 and Q_1 are the first terms in the Higgs factoring of the strong maps σ and τ.

The matroid join describes the Higgs factoring of $F(S) \to M(S)$. We have already seen that $N \vee U_1(S)$ is the free-est matroid on S which has its rank equal to $\text{rk}\, N + 1$ and N as a strong image. Thus $N \vee U_1(S)$ is the penultimate term in the sequence. The other terms are given by $N \vee U_k(S)$ for $0 \leq k \leq |S| - \text{rk}\, N$.

Looking back at the matroid join construction from the point of view of maps, we see that there are strong maps

$$M_1 \vee M_2 \longrightarrow M_1$$
$$M_1 \vee M_2 \longrightarrow M_2$$

formed by contracting S_2 and S_1 from the full matroid on $S \cup S_1 \cup S_2$. However $M_1 \vee M_2$ is certainly not the least free such matroid as can be seen by putting $M_1 = M_2$, nor is it the free-est as the diagram shows.

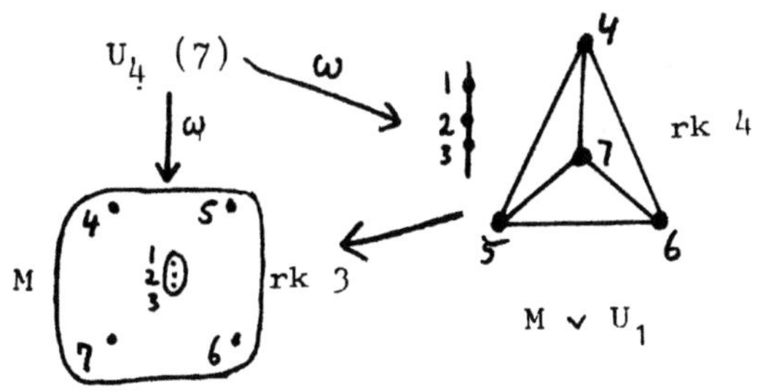

Figure 26 $M \vee U_1$ is not the free-est

This suggests two problems for investigation:

1. Find necessary and sufficient conditions for $N(S) = M_1 \vee M_2(S)$ non trivially.
2. Characterize $M_1 \vee M_2$ in terms of strong and weak maps.

Truncating and Erecting

Another example of strong and weak maps working together can be found in Crapo's work [2d] on erecting geometries. Recall that M^T is formed by making all bases into circuits, or more geometrically, adding a point freely on the flat S of M and then contracting it out. Crapo showed that for each N(S) the collection of all matroids M on S such that $M^T = N$ forms a lattice in the weak ordering. In particular there is a free-est such M. It is not true however, for matroids M such that $M^{TT} = N$.

A detailed study of the Higgs lift and the Crapo erecting led Las Vergnas to more general results [9], of the following type:

Theorem Suppose that we have two strong maps of degree one as shown: ($\text{rk } M_i - \text{rk } M_0 = 1$)

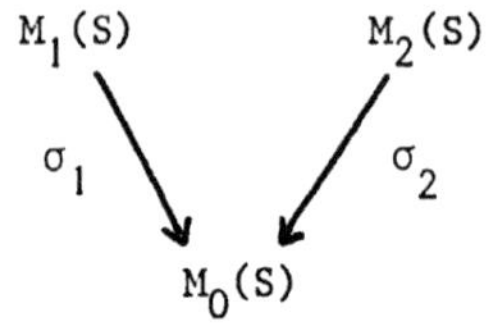

Then there exists P(S) free-er than both M_1 and M_2 such that

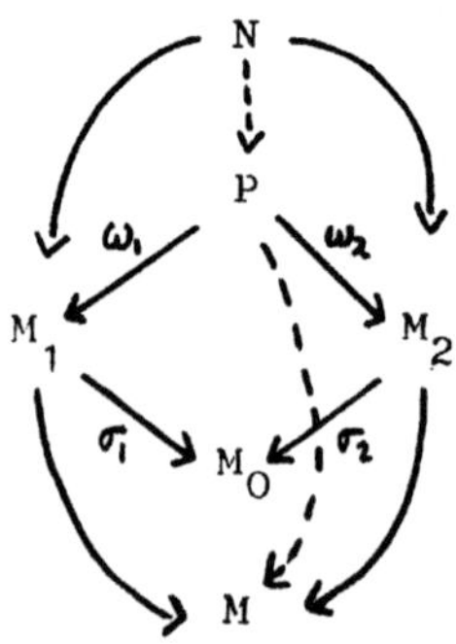

$$\text{rk } M_1 - \text{rk } M_0 = 1$$
$$\text{rk } M_2 - \text{rk } M_0 = 1$$

(a) $M_1 \rightarrow M$ and $M_2 \rightarrow M$
$\Rightarrow P \rightarrow M$ (all strong maps)

(b) $N \rightarrow M_1$ and $N \rightarrow M_2$
$\Rightarrow N \rightarrow P$ (all strong maps)

However this canonical P may not exist if $\text{rk } M_1 - \text{rk} M_0 > 1$. These investigations are really concerned with the weak partial order of matroids on a set S, and demonstrate that often the matroids of the same rank which satisfy some diagram form a lattice, but not when the ranks are allowed to vary.

Algebraically the definition of a strong map makes sense since the inverse image of a subvector space is a subvector space. Higgs, following an idea of Edmonds, was able to show that every strong map can be considered as an extension followed by a contraction, thus linking approaches (i) and (ii) in the search for maps. I shall discuss his result in detail shortly.

Unfortunately, the only standard categorical ideas which work are the free object on S, and the coproduct (matroid sum M + N).

1.6 A Geometric Picture of Duality

We are now in a position to investigate matroid duality more closely. A good place to begin is with Whitney's work [19] on representable matroids. Recall that a matroid M(S) without loops is representable in a vector space V over K if there is a one-one function

$$f : S \longrightarrow V$$

$$A \text{ indep in } M \Longleftrightarrow f(A) \text{ linearly indep in } V.$$

Suppose M is represented in a vector space of dimension r = rk M. Form the matrix of column vectors $f(s) : s \in S$.

f(1)	f(2)	f(3)
1	1	0
0	0	1
0	1	-1

Now look at a new space $W = K^{|S|}$ spanned by basis vectors $b_s : s \in S$. The rows of the matrix span an r-dimensional subspace H of W. For any subset T of S let Π_T denote the projection of W onto the subspace spanned by $\{f(t) : t \in T\}$. Then $\rho T = \dim \Pi_T(H)$. The dual of M is formed analogously. Whitney takes $K = \mathbb{R}$ and so talks about the subspace $H^\perp$ orthogonal to H and discovers that $\rho^*T = \dim \Pi_T(H^\perp)$. By changing perspective slightly we can remove the necessity for orthogonality and get a construction which generalizes to arbitrary matroids. Instead of working with $\Pi_T(H)$, let Π_H denote the projection onto the subspace spanned by H. Then $\dim \Pi_T(H) = \dim \Pi_H(T)$ (second isomorphism theorem). Now take any subspace H^* complementary to H in W. Then

$$\rho T = \dim \Pi_H(T)$$

$$\text{and } \rho^*T = \dim \Pi_{H^*}(T)$$

In the example, $H^* = \langle(1,-1,-1)\rangle$. Projection onto H^* gives a rank one matroid with each point being independent, thus illustrating that $U_2^*(3) = U_1(3)$.

The essence of Whitney's construction is a representation of the free matroid F(S). Inside the linear span of F(S) he finds two complementary subspaces which produce M and M* by projection. But the matroid join construction does all of this except the "linear" part!

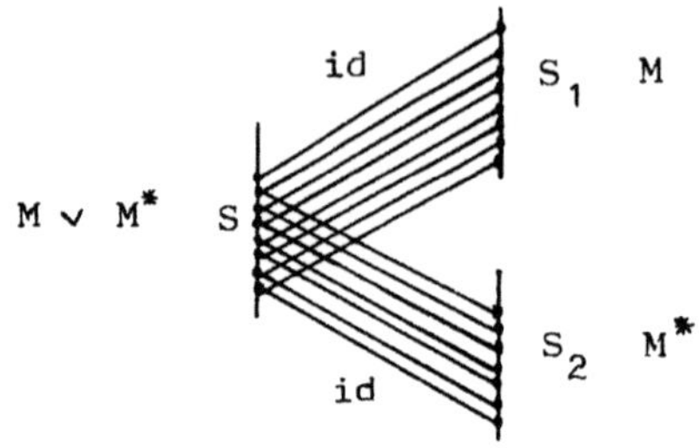

Think of the full matroid $M \vee M^*$ on $S_1 \cup S_2 \cup S$. Then S is a basis for $M \vee M^*$ since bases of M and M^* are complementary. Thus $M \vee M^*|S$ is free. Also $M(S_1)$ and $M^*(S_2)$ are complementary and

$$(M \vee M^*)/S_1 \text{ is isomorphic to } M^*$$

$$(M \vee M^*)/S_2 \text{ is isomorphic to } M.$$

When discussing the Higgs lift we saw that M^* could be replaced by $U_d(S)$ where $d = |S| - rkM = rkM^*$. In fact any matroid on S which shares one basis with M^* will also do. Yet intuitively there is something canonical about using M^*. The best that I can offer is that M^* is the unique matroid on S with the property that for every N with $M \vee N = F$, some base of M^* is indep in N.

I shall conclude this lecture with a brief introduction to the second. During the preparation of these notes, I have been struck by how useful it is when confronted by a matroid construction, to encompass the original matroids (the pieces) and the final construct, in one large matroid. Strong maps are no exception.

Suppose the identity $S \rightarrow S$ is a strong map of $M \rightarrow N$. Let us take $rkM - rkN = 1$ for simplicity. Let S_1 be an identical copy of S. I wish to construct a matroid P on $S \cup S_1 \cup \{p\}$ such that contraction by p projects M onto N.

Getting started is easy, since I can adjoin p to M by its modular cut in the Higgs factoring. Now all that is needed is to intersect each of the lines <p,s> : $s \in S$ with a hyperplane $N(S_1)$. This is exactly what the Dilworth Truncation does.

2. DILWORTH TRUNCATION AND GENERALIZATIONS

Graph theorists have vertices and edges to play with. When we pass to the cycle matroid we carry over only the structure of the edges of the graph. Sometimes, by subtle argument, it is possible to convert graphical ideas which seem to demand both vertices and edges, into statements depending on edges only and hence into matroids. Tutte is a master of this, having made the transition for vertex colourings and for Menger's theorem [16b]. However it still leaves us with the problem of understanding geometrically what is happening.

Working with vector spaces we have a coordinate structure which governs the nature of linear maps. When we pass to matroids, we lose this structure. One result is that matroid maps applied to a vector space matroid are not all linear. Although the power of matroids is the removal of the coordinate structures, we are left with a more general structure which fails to be categorically nice - at least in the attempts made so far.

With these thoughts in mind we encounter the idea of a comap due to Crapo [2c], which generalizes a construction of Dilworth. At present, very little is known about the structure of comaps, so I shall concentrate mostly on Dilworth's truncation construction. It turns out that it is related to matroid join and has relevance in the problem of deciding when a matroid is representable. Furthermore, a clever construction due to Dowling [5] which can be viewed as a comap construction, suggests a way of co-ordinatizing a matroid which may give a useful category of maps mimicking linear transformations more closely than do strong maps.

2.1 The Dilworth Truncation

Dilworth was interested in problems of embedding lattices in semi-modular lattices with certain properties [4], and was led to the construction on which I wish to concentrate. His idea was to take the flats which have rank k, and make them the atoms of a new geometric lattice by introducing new flats in such a way that old flats of rank k + d become flats of rank d in the new geometry.

Geometrically the idea is very simple. Think of the matroid as displayed in space and place a new hyperplane in general position meeting every line of the matroid M.

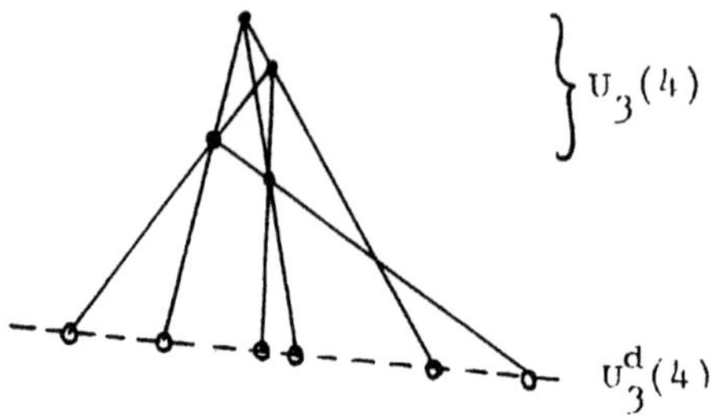

Figure 1 Dilworth truncation of $U_3(4)$. See also Figure 4.

The new points on the hyperplane form the Dilworth lower truncation M^d of M. Each line of M corresponds to a new point of M^d, each old plane to a new line and so on.

The illustrations indicate what happens when lines of M become points of M^d. More generally, Dilworth described how to make the k-flats of M into points of a new geometry $M^{d(k)}$. From the requirement that old flats of M remain as flats of the Dilworth truncation (corresponding to the requirement that the lower truncated lattice of M be lattice embedded in the new geometry), it follows that $M^{d(2)}$ is a restriction of M^{dd}.

Let us look in detail at M^d. When will three old lines be dependent in M^d? Precisely when they are coplanar in M. That is, when, as a set of points of M, they span a flat of rank 3.

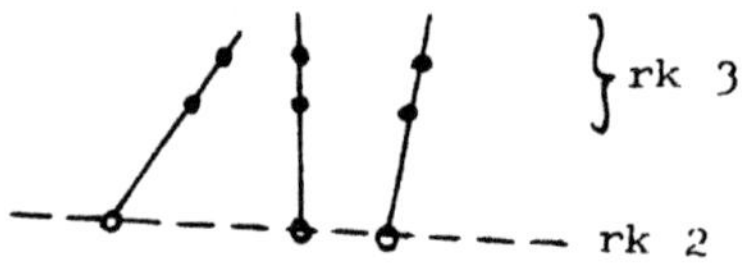

Figure 2 Truncating coplanar lines

This sort of consideration leads directly to Dilworth's construction.

Denote by $\bar{L}_i$ the new point to be placed on the old line L_i of M. Then M^d is specified by

$$\{\bar{L}_i, \ldots, \bar{L}_t\} \text{ is indep in } M^d \Leftrightarrow \text{ For any } J \subseteq \{1, \ldots, t\} \quad \operatorname{rk} \bigcup_J L_j \geq |J| + 1$$

I shall defer checking that this is a bonafide matroid until later when we can consider it in a more general context. Instead let us look at what happens in M^d.

(i) Suppose $\{b_1,\ldots,b_n\}$ is a basis of M. Let $L_i = \langle b_1,b_i\rangle$. Then $\{\bar{L}_i : 2 \le i \le n\}$ is a basis of M^d.

Proof: For any $J \le \{2,\ldots,n\}$ we have

$$\mathrm{rk} \bigcup_J L_j = |J| + 1$$

so the set is independent. Further, if L is any other line of M, then

$$\mathrm{rk}\ L \cup \bigcup_2^n L_j = n < n + 1$$

so $\bar{L}$ is in the span of $\{\bar{L}_j : 2 \le j \le n\}$

(ii) The contraction of M by the point x is the restriction of M^d to the points $\langle\overline{x,s}\rangle : s \in S\backslash\{x\}$

$$M/x = M^d | \{\langle\overline{x,s}\rangle : s \in S\backslash\{x\}\}$$

Proof: $\{\langle\overline{x,b}\rangle : b \in B\}$ is indep in M^d
$\Leftrightarrow$ $\mathrm{rk}\{x\} \cup A \ge |A| + 1$ for all $A \le B$
$\Leftrightarrow$ $B \cup x$ is indep in M
$\Leftrightarrow$ B is indep in M/x.

(iii) It follows from (ii) that $(M+e)^d$ contains both M and M^d, in the sense that $(M+e)^d$ encompasses both M and the geometric construction of M^d from M.

(iv) As Dilworth pointed out, F_n^d is the cycle matroid of K_n whose lattice of flats is the partition lattice of an n-element set.

Proof: Each point $\bar{L}$ of F_n^d corresponds to the line L of F_n, which can be thought of as an edge of K_n joining the two points of F_n which form L. A set $\{\bar{L}_1,\ldots,\bar{L}_t\}$ of points in F^d is indep

$\Leftrightarrow$ $\mathrm{rk}\ L_j \ge |J| + 1$ for all $J \le \{1,\ldots,t\}$
$\Leftrightarrow$ every subset of J edges meets at least $|J| + 1$ points
$\Leftrightarrow$ the edges $L_1,\ldots,L_t$ form a forest in K_n.

Since each flat of the cycle matroid of K_n is a union of vertex disjoint complete subgraphs of K_n, it corresponds precisely to a partition of the set of n vertices. The rank of the flat is n minus the number of blocks of the partition.

As an example, consider the case $F_5^d = K_5$. The graph K_5 is familiar, but how many people recognize it as the Desargues Configuration? Since $F_4 + e = F_5$ we can construct F_5^d by looking at the construction of F_4^d from F_4, to get the diagram shown below.

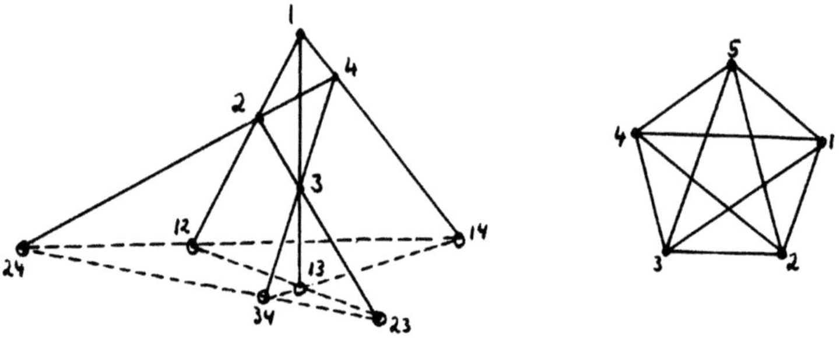

Figure 3 Desargues Configuration as K_5

This is geometrically the Desargues configuration and at the same time it is the cycle matroid of K_5. Now we can translate the perspectivity of two triangles back into the graph. The points of the triangles correspond to edges of the graph not forming a cycle. The three lines of perspectivity are cycles in K_5, so taking edge 15 as the point 1 of perspectivity, the two triangles 234 and 12,13,14 correspond to the two stars 25,35,45 and 12,13,14, which together with edge 15 form three 3-cycles. The remaining edges 24,34,23 lie on a line in F_5^d and correspondingly form a 3-cycle in K_5.

Recalling the maxim that whenever possible with a matroid construction it is worthwhile trying to encompass both M and the result in one large matroid, we observe that not only does the Dilworth Truncation do this automatically ($(M+e)^d$ is exactly this for M and M^d), it also answers the question at the end of the last lecture. Namely, given $x \in S$ then the construction of M/x from M is already sitting inside $(M+e)^d$ as a restriction. By looking at F_5^d, that is F_4 and F_4^d together, several ideas emerge which are worth pursuing.

(i) Why not put points on other flats as well, not just one level at a time as Dilworth does, but simultaneously? (Section 2.2)

(ii) Putting points freely on lines suggests links with the matroid join. (Section 2.3)

(iii) The Desargues configuration appears whenever we look at four independent points in M and the configuration they generate in $(M+e)^d$. Since Desargues theorem is the basis for linear representation of projective geometries, perhaps there is some connection with representability. (Section 2.4)

(iv) The hyperplane was taken to be as free as possible. By making it less free, we get more general comaps. Geometrically this ideas is straightforward, at least conceptually. Think of M as a configuration in space, and let a hyperplane move about, always intersecting all the lines of M. The results are certain weak map images of M^d. It would be very useful indeed to have some manageable algebraic method of doing this.

2.2 Generalization of Dilworth Truncation

The specification of M^d (granted that we haven't yet proved it is a matroid) suggests that it could be generalized to other flats of M as follows. To each flat F of M, associate a new point $\bar{F}$. It is convenient for what follows to exclude $\langle\phi\rangle$.

$$\text{Specify } \{\bar{F}_1,\ldots,\bar{F}_t\} \text{ indep} \Leftrightarrow \text{For any } J \subseteq \{1,\ldots,t\} \quad \text{rk} \bigcup_J F_j \geq |J| + 1$$

Ignoring for a moment the question of whether this is a matroid, let us look at the effect. Let $x \in S$. Then $\langle x\rangle$ is dependent, since $\text{rk}\{x\} = 1 < 1 + 1$. This could be coped with by altering the additive constant 1. In fact, if we look only at flats of rank k in M, then specifying

$$\text{rk} \bigcup_J F_j \geq |J| + k-1$$

actually produces the k-th Dilworth truncation, turning $\bar{F}$ into an independent point and having overall rank equal to rk M − k + 1. For k = 1 we of course get M back again. We shall return to the case k = 1 later when we recognize it as a generalization of matroid joins.

Thus by adjusting k we can achieve the various Dilworth truncations. To get them all simultaneously we need to adjust k, depending on the rank of the flats we are considering. Furthermore, if $\{b_1,\ldots,b_t\}$ is an indep set in M and F is their span, we would like $\{\bar{b}_1,\ldots,\bar{b}_t,\bar{F}\}$ to be independent in the new matroid. Putting all this together suggests the following general construction M^D.

Given a matroid M(S), take as a new set of points

$$\{\bar{F} : F \text{ a flat of } M \text{ with rank} \geq 1\}$$

Put $\{\bar{F}_1,\ldots,\bar{F}_t\}$ indep in $M^D \Leftrightarrow$ For any $J \leq \{1,\ldots,t\}$

$$\text{rk} \bigcup_J F_j \geq |J| + \underset{J}{\text{Min}} \{\text{rk } F_j\} - 1$$

Before proving that M^D really is a matroid, it is worthwhile looking in detail at what is happening.

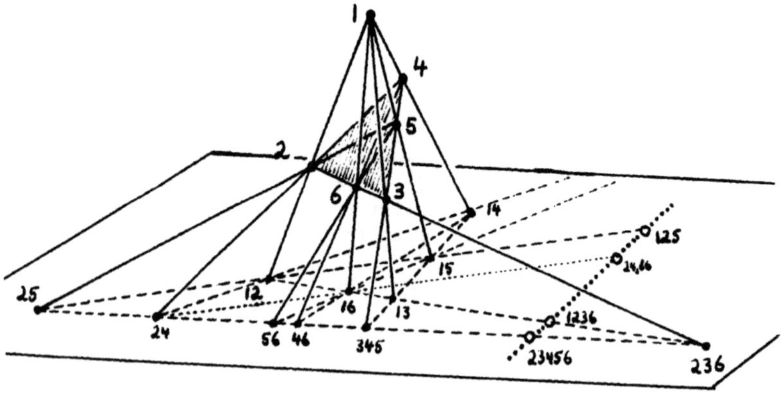

Figure 4 A typical M^D (not all points are shown)

(i) Each point $\bar{F}$ is indep in M^D since interpreting the condition when J is a singleton gives

$$\text{rk } F \geq 1 + \text{rk } F - 1$$

(ii) For any $A \leq S$, A indep in $M \Leftrightarrow \{\bar{a} : a \in A\}$ indep in M^D. Here the Min is 1 so

A indep in $M^D \Leftrightarrow$ for any $B \leq A$ rk $B \geq |B| + 1 - 1$
A indep in M.

(iii) Let $\tilde{\mathfrak{F}}_k = \{\bar{F} : \text{rk } F = k\}$. Then M^D restricted to $\tilde{\mathfrak{F}}_k$ gives the k-th Dilworth truncation of M which has rank equal to rk M - k + 1.

(iv) The point $\bar{F}$ lies freely on $M|(F \cup \{\bar{F}\})$. For any basis $B = \{b_1, \ldots, b_t\}$ of F in M, we find that $\{\bar{b}_1, \ldots, \bar{b}_t, \bar{F}\}$ is dependent in M^D but every proper subset is independent.

(v) $\bar{F}$ lies freely on the line $<\bar{G}_1, \ldots, \bar{G}_t : G_i \leq F, \text{rk } G_i = \text{rk}F - 1>$.

(vi) If instead of forming M^D directly we build a matroid recursively by

$$M_0 = M$$

$$M_{n+1} = (M_n + e_{n+1})^d$$

then $M_{N+1} = M_N$ for all $N \geq$ rk M and we get a larger matroid which contains M^D. This larger matroid is supersolvable - that is it has a chain of modular flats and this has ramifications on the chromatic and Whitney polynomials [15].

Theorem M^D is a matroid.

Proof Let the set $\tilde{\mathfrak{F}} = \{\bar{F} : \text{rk } F \geq 1\}$ be enumerated as $F_1, \ldots, F_N$. For any $J \leq \{1, \ldots, N\}$, put $\lambda(J) = \text{rk} \bigcup_J F_j - \min_J \text{rk}F_j + 1$

Then M^D is specified by

$\{\bar{F}_i : i \in I\}$ is indep in $M^D \Leftrightarrow J \leq I, \lambda(J) \geq |J|$.

This function λ is increasing and submodular (since the rank function of M is):

for $J \leq K$, $\lambda(J) \leq \lambda(K)$
for arbitrary J and K, $\lambda(J) + \lambda(K) \geq \lambda(J \cup K) + \lambda(J \cap K)$.

Thus λ looks very much like a rank function. However $\lambda(\phi) = 1$ and $\lambda(J \cup \{k\})$ may be greater than $\lambda(J) + 1$. There are standard ways to convert λ to a rank function, but a quick way to see that we do have a genuine matroid is as follows.

Let C_1 and C_2 be distinct sets, minimal with respect to not being independent, and let $e \in C_1 \cap C_2$. We show that $C_1 \cup C_2 \setminus \{e\}$ is not independent, thus verifying the circuit exchange axiom for a matroid.

Since C_1 and C_2 are minimal-not-independent, we find:

$$|C_i| > \lambda C_i \geq \lambda(C_i \backslash \{e\}) \geq |C_i| - 1$$

Therefore $\lambda(C_1 \cup C_2\backslash\{e\}) \leq \lambda(C_1 \cup C_2)$ since λ is increasing

$\leq \lambda(C_1) + \lambda(C_2) - \lambda(C_1 \cap C_2)$ since λ is submodular

$\leq |C_1| - 1 + |C_2| - 1 - |C_1 \cap C_2|$ since $C_1 \cap C_2$ is indep

$= |C_1 \cup C_2| - 2$

$< |C_1 \cup C_2\backslash\{e\}|$.

Thus $C_1 \cup C_2\backslash\{e\}$ is not independent, hence M^D is a matroid.

The reason for including this well known argument is to permit me to mention an open question concerning such functions λ, as well as illustrating how choosing the aspect carefully (in this case circuits instead of rank) can simplify the arguments considerably. (The usual rank proof is much more tedious.)

The proof depended only on λ being increasing and submodular. Other examples of such functions are Max,-Min, $\lambda_1 + \lambda_2$, $\alpha\lambda$ for $\alpha \geq 1$, constant. The collection of all such λ's on a set S form a cone under positive linear combinations. I believe it is still an open question to describe the extremal members of this cone.[1] In this connection Murty, observing that λ is usually different from the rank function, has asked for a description of those matroids for which the only such λ is the rank function.

2.3 Dilworth Truncations and Matroid Joins.

The matroid join $M_1 \vee M_2$ of $M_1(S)$ and $M_2(S)$ can be looked at as a matroid on three copies of S: S, S_1, S_2. It can be described geometrically as putting the points of S freely on the corresponding lines $\langle s_1, s_2 \rangle$ of $M_1(S_1) + M_2(S_2)$. The Dilworth truncation of $M_1 + M_2$ puts points freely on all the lines of $M_1 + M_2$ but confines them to a hyperplane. The connection between the two is that by adjusting the additive constant in the Dilworth truncation, the specification

$$\{\bar{F}_1, \ldots, \bar{F}_t\} \text{ indep} \Leftrightarrow \text{For any } J \subseteq \{1, \ldots, t\} \quad \text{rk } F_j \geq |J|$$

when applied to $M_1 + M_2$, contains $M_1 \vee M_2$ as a restriction. The usual statement of independence for $M_1 \vee M_2$ is

1. This has been answered by Alan Cheuny and Henry Crapo in an article concerning Quotient Bundles to appear soon in J.C.T.B.

(i) $A \leq S$ indep $\Leftrightarrow$ $\exists\, B \leq A', \rho_1 B + \rho_2 A\backslash B \geq |A|$

Our condition translates into

(ii) $A \leq S$ indep $\Leftrightarrow$ $\forall\, B \leq A, \rho_1 B + \rho_2 B \geq |B|$.

The only proofs known to me of the fact that these are equivalent are quite difficult and technical matroid arguments and inappropriate for inclusion here. Statement (ii) is the version that Edmonds first used, statement (i) comes from Nash Williams.

2.4 Representability of Dilworth Truncations

We have seen that the Desargues configuration is involved in F_4^D, and that F_4 is a submatroid in many ways of any matroid M of rank 4. Since Desargues theorem is the configuration behind linear representation of projective geometries, it seems natural to look at representability of M, M^d and M^D.

Let me remind you that a matroid M(S) (without loops) is linearly representable over characteristic χ if and only if there exists a one-one function $f : S \to V$ where V is a vector-space over a field of characteristic χ, such that

A indep in M(S) $\Leftrightarrow$ f(A) linearly indep in V.

<u>Theorem</u> If rk M ≥ 4 then M representable over χ $\Leftrightarrow$ M^d representable over χ.

<u>Remark:</u> It follows that the same result holds for M^D since M^D is a restriction of a sequence of $(M+e)^d$ constructions.

<u>Proof outline</u>

M representable $\Rightarrow$ M^d representable.

Pick a representation of M with basis $b_1,\ldots,b_n$ of M represented as $\phi(b_i) = \underline{b}_i$ in V. Each rk one flat of M is represented as a one-dimensional subspace of V. Extend the field K to $K(\alpha_2,\ldots,\alpha_n)$ where the α's are algebraically independent. Let H be the hyperplane of V spanned by $\{\underline{b}_1 + \alpha_i \underline{b}_i : 2 \leq i \leq n\}$. The α's were chosen to be algebraically independent to keep H well away from M, and so that each line L of M, (which corresponds to a two dimensional subspace of V) meets H in a <u>unique</u> one dimensional subspace of V. Then the representation $\phi(\mathfrak{L})$ of $\mathfrak{L}$ is any non-zero vector in this subspace.

We must also show that ϕ does give a representation of M. The algebraic details are hiding in the uniqueness property of H mentioned above.

Suppose $\bar{L}_1,\ldots,\bar{L}_t$ are independent in M^d.

Then rk $\bigcup_J L_j \geq |J| + 1$ for all $J \leq \{1,\ldots,t\}$. Therefore the subspace W_J of V spanned by $\{\phi(K_j) : j \in J\}$ must have dimension

$\geq |J| + 1$ in V.

Thus dim $W_J \cap H \geq |J|$. Consequently the vectors

$\{\phi(\bar{L}_1),\ldots, \phi(\bar{L}_t)\}$

are linearly independent in V.

Conversely, if

$\{\phi(\bar{L}_1),\ldots, \phi(\bar{L}_t)\}$

are linearly independent in V then dim $W_J \cap H \geq |J|$ and since no L_i is contained in H,

dim $W_J \geq |J| + 1$.

Thus rk $\bigcup_J L_i \geq |J| + 1$ whence $\{\bar{L}_1,\ldots,\bar{L}_t\}$ is independent in M^d.//

M^d representable $\Rightarrow$ M representable (rank M ≥ 4).

Assume M^d is representable in V over the field K. Choose a basis $e_1,\ldots,e_n$ of M, and choose a basis $\underline{b}_1,\ldots,\underline{b}_n$ of V over K. Extend K to $K(\alpha_2,\ldots,\alpha_n)$ as before. Then $\{\underline{b}_1 + \alpha_i\underline{b}_i : 2 \leq i \leq n\}$ are linearly independent and so can be chosen to represent the basis $\{\overline{\langle e_1,e_i\rangle} : 2 \leq i \leq n\}$ of M^d. This can be extended to a representation of M^d with coefficients from K, relative to this basis. We can also arrange our representation of M^d so that

$\overline{\langle e_2,e_i\rangle}$ is represented as $(\underline{b}_1 + \alpha_2\underline{b}_2) + (\underline{b}_1 + \alpha_i\underline{b}_i)$.

Using the fact that

$\overline{\langle e_1,e_i\rangle}$, $\overline{\langle e_1,e_j\rangle}$, $\overline{\langle e_i,e_j\rangle}$

and

$\overline{\langle e_2,e_i\rangle}$, $\overline{\langle e_2,e_j\rangle}$, $\overline{\langle e_i,e_j\rangle}$

must de distinct lines in V, the representation of $\overline{\langle e_i,e_j\rangle}$ must be $(\underline{b}_1 + \alpha_i\underline{b}_i) - (\underline{b}_1 + \alpha_j\underline{b}_j)$ for $2 < i < j$.

We are now ready to represent M. The vectors $\underline{b}_1,\ldots,\underline{b}_n$ have been constructed so that they represent $e_1,\ldots,e_n$ corresponding to our representation of M^d. We proceed inductively. Consider any element e in S not yet represented. For each element x already represented, we know the co-ordinates of $\overline{\langle x,e\rangle}$ in M^d. It is easy enough to deduce the co-ordinates of e, but we must also consider the embarrassment of all the extra information - is it consistent?

Since $rkM \geq 4$, there are three base elements e_i,e_j,e_k such that e does not lie on any of the lines they generate. For definiteness, suppose it is e_1,e_2,e_3. (Note: the special role of b_1 is apparent, not real.) Then the known co-ordinates of $\overline{\langle e,e_2\rangle}$ and $\overline{\langle e,e_3\rangle}$ are sufficient to determine co-ordinates for e, and all the other information $\overline{\langle x,e\rangle}$ is consistent. The co-ordinate computation is straightforward. You end up with 2n equations in n+3 unknowns, but they are seen to be consistent by observing the consequences of the fact that $\overline{\langle e_2,e\rangle}$, $\overline{\langle e_3,e\rangle}$, $\overline{\langle e_2,e_3\rangle}$ lie on a line in V. Similar arguments show that the co-ordinates of $\overline{\langle e_1,e\rangle}$ are consistent with M^d.

The more general consistency problem is treated as follows. Suppose that we wish to check $\overline{\langle x,e\rangle}$. Since $\{e_1,e_2,e_3\}$ is independent in M, we can find two base elements e_i and e_j $(1 \leq i < j \leq 3)$ so that $\{x,e,e_i,e_j\}$ are independent in M. Then the two lines

$$\overline{\langle e_i,e\rangle},\ \overline{\langle e_i,x\rangle},\ \overline{\langle e,x\rangle}$$

and

$$\overline{\langle e_j,e\rangle},\ \overline{\langle e_j,x\rangle},\ \overline{\langle e,x\rangle}$$

are distinct, and the co-ordinates of $\overline{\langle e_i,e\rangle}$, $\overline{\langle e_i,x\rangle}$, $\overline{\langle e_j,e\rangle}$ and $\overline{\langle e_j,x\rangle}$ are already known to be consistent with M^d. The four points e,e_i,e_j,x give rise to a Desargues configuration in M with M^d, and so in V. Thus the co-ordinates of $\overline{\langle e,x\rangle}$ are uniquely determined and consistent with M^d.

The importance of Dilworth truncation for deciding if a matroid is representable is due to the existence of Desargues configurations. For example, the smallest (in size) non-representable matroid is V_4 (or one of its many weak images). By constructing V_4^d and looking closely, one discerns a rank 3 Desargues$^-$ configuration in V_4^d showing again that V_4 cannot be linearly represented.

The Vamos matroid shows up in other contexts too. This example is due to Ingleton.

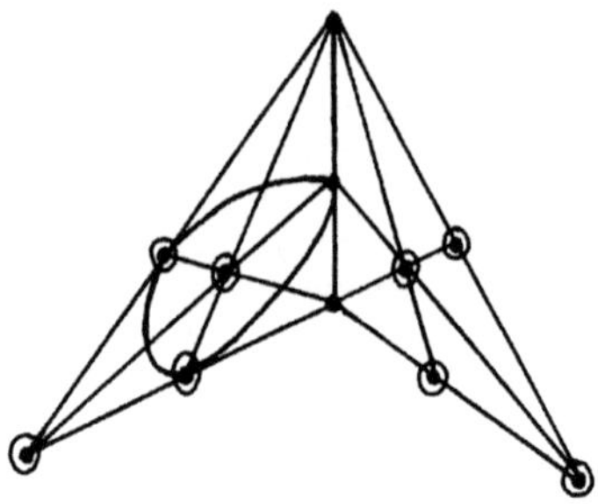

Figure 5 A non-representable matroid of rank 4

We know from algebraic considerations that one plane requires a field of characteristic 2 (because of the Fano configuration) and the other plane demands characteristic not 2. Hence it is not representable. The circled points form a V_4 so we can see geometrically why this matroid is not representable.

More generally, if a matroid is representable, then we can use the ideas of projectivities to observe that any basis together with a free point should be projective with any other such set. By this I mean that it is possible to construct a new matroid from the old, with a new point collinear with the chosen pairs of projective points, thus constructing the projectivity as an explicit perspectivity. With many (but not all) non-representable matroids, an attempt to perform this construction either runs into a blatant impossibility or else contains a Vamos configuration. In either case this shows the matroid to be non-representable.

2.5 The Lattice Approach

We have seen that the ordinary Dilworth truncation of the free matroid F_n produces the cycle matroid of K_n whose lattice of flats is the lattice of partitions of an n-element set. This idea carries over more generally to M^d.

Let M(S) be a matroid, L the lattice of flats of M, and A the set of atoms of L (flats of rank one in M). For any subset X of A we denote by $\delta(A\backslash X)$ the discrete partition of $A\backslash X$. To each element x of L, we associate the set X of atoms below x. We call π an <u>L-partition</u> of A if π is a partition of A whose blocks are elements of L, and a <u>partial L-partition</u> of A is a partition of some subset X of A corresponding to an element of L, all of whose blocks are also elements of L. For $M = F_n$, L is the boolean on S and the L-partitions become the flats of K_n, while the partial L-partitions become the flats of K_{n+1}, that is, of $(F_n+e)^d$.

$$\begin{array}{rcl} M & \leftrightarrow & L \\ M^d & \leftrightarrow & \text{L-partitions} \\ \bar{F} & \leftrightarrow & [F,\ \delta(A\backslash F)] \\ (M+e)^d & \leftrightarrow & \text{partial L-partitions.} \end{array}$$

It turns out that the L-partitions of A, and the partial L-partitions of A are geometric lattices - in fact they are just the lattices of flats of M^d and $(M+e)^d$ respectively. The lattice of L-partitions of M+e gives rise to the partial L-partitions of M by deleting all blocks containing e. The ordering and lattice operation are:

$$\pi_1 \le \pi_2 \Leftrightarrow \bigcup\pi_1 \le \bigcup\pi_2 \text{ and each block of } \pi_2 \text{ is a union of blocks of } \pi_1$$

$$\pi_1 \vee \pi_2 = \text{common refinement of } \pi_1 \text{ and } \pi_2.$$

The reason for going to this level of detail is that the idea of partial partitions is useful, not only in describing Dilworth truncation, but also in transversal theory [11]. Dowling also exploited this lattice in a paper [5] in which he constructs matroids which are partly co-ordinatized by elements of a group.

Basically, Dowling's construction is to take a group G and a set S, and replace the one-flats s of F(S) by the set of parallel elements $\{gs : g \in G\}$. This produces a matroid on $\{gs : g \in G,\ s \in S\}$ whose only circuits are two element sets of the form $\{gs, hs\}$.

Now lift this matroid one rank to get the matroid illustrated, with rank $|S| + 1$.

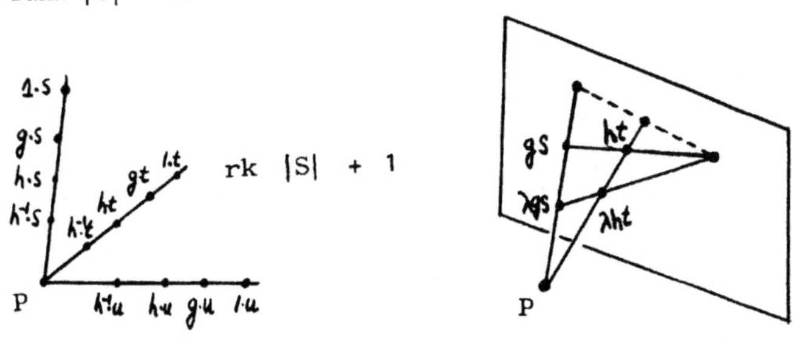

Figure 6 Dowling's Construction

The point p is acting like an origin, with the G-multiples of S stretched out along the direction of s. Now form a kind of Dilworth truncation, but identify

$$\langle gs, ht\rangle \text{ with } \langle \lambda gs, \lambda ht\rangle$$

for all $\lambda \in G$. This is a particular example of a comap. Contracting by p (so that it really acts like the origin) produces Dowling's matroid. The group structure is sufficiently embedded in this matroid to force any linear representation to have G as a subgroup of the multiplicative group of the field.

It should be possible to replace F(S) by an arbitrary matroid M(S). However, a much more general context should also be obtainable along the following lines.

2.6 Generalized Coordinates

Take the free matroid F(S). (Later we can replace F by an arbitrary matroid on S.) To each $s \in S$, assign a distinct set K_s and consider the matroid N on K_s in which K_s is the one-flat previously called s in F(S). (This works for any matroid.) We now have N looking rather like the axes of a vectorspace with K_s the scalar multiples of s. Let p be a new point, and lift N by p to give N_p. We could now form N_p^D which corresponds in vector spaces (as far as I can tell) to taking K_s to be algebraically independent over a ground field K.

When we generate a vectorspace by saying "closed under linear combinations" we are taking a weak image of N_p^D in which new points corresponding to old lines with proportional coefficients are identified.

Dowling took K_s to be a group G and showed how to construct the weak image when the new points are identified if their co-ordinates are proportional by left multiplication in G. What would be interesting would be to classify just which equivalence relations on the K_s will give rise to a matroid. This is the same thing as classifying the weak images of N_p^D - a hard problem.

We can take this idea one step further. I suspect that it would be useful to say that a matroid M is coordinatized by a matroid N_p if some weak image of N_p^D contains M as a restriction. This would truly mimic vector spaces since p is acting as the origin. Now we can introduce a co-ordinate map. For example M_1 can be mapped to M_2 with respect to a co-ordinate system N_p if M_2 can be obtained (up to isomorphism) by contractions (projections) by points inside N_p. This would generalize more closely the idea of a linear transformation than does the very general strong map.

References

1. T. Brylawski An Outline for the Study of Combinatorial PreGeometries. Notes, Univ. of N. Carolina 1972

2. H. Crapo (a) Structure Theory for Geometric Lattices. Rend. Sem. Mat. Univ. Padova 38(1967)14-22.
(b) Single Element Extensions of Matroids. J.Res.Nat.Bur.Standards 69B (1965)55-65.
(c) The Joining of Exchange Geometries. J.Math.Mech. 17(1968)837-852 MR 39#2657.
(d) Erecting Geometries. Proc.2nd Chapel Hill Conf. on Comb.Math and Applications, 74-79.

3. H.Crapo - G.C. Rota On the Foundations of Combinatorial Theory: Combinatorial Geometries M.I.T. Press, Camb. Mass. 1970.

4. R. Dilworth Dependence Relations in a Semimodular Lattice. Duke Math. J. 11(1944)575-587.

5. T. Dowling A Class of Geometric Lattices Based on Finite Groups J.C.T. 14(1)1973, 61-86.

6. J. Edmonds - D. Fulkerson Transversals and matroid Partition. J.Res.Nat.Bur.Standards 69B(1965) 147-153.

7. D. Higgs (a) Strong Maps of Geometries. J.Comb.Th. 5(2)1968, 185-191.
(b) Maps of Geometries. J. Lond. Math. Soc. 41(1966)612-618.

8. A. Ingleton Conditions for Representability and Transversality of Matroids. Theorie des Matroides, Lecture Notes 211 Springer Verlag 1971, 62-66.

9. M. Las Vergnas On Certain Constructions for Matroids. Proc. 5th British Comb.Conf.1975,395-404.

10. S. Maclane Some Interpretations of Abstract Linear Dependence in terms of Projective Geometry. Amer. J. Math. 28(1937)22-32.

11. J. Mason The Three Family Problem. Open Univ. Seminar Notes 1976.

12. L. Mirksy Transversal Theory, Academic Press 1971.

13. C. Nash-Williams An Application of Matroids to Graph Theory. Proc. Symp. Rome, Dunod 1968, 263-265.

14. R. Rado A note on Independence Functions. Proc. Lond. Math. Soc. 7(1957)300-320.

15. R. Stanley Super Solvable Semimodular Lattices Proc. Conf. on Mobius Algebras, Univ.Waterloo 1971, 80-142.

16. W. Tutte (a) Introduction to the Theory of Matroids. Rand Report 448-Pr 1966.
(b) Menger's Theorem for Matroids. J.Res.Nat.Bur.Standards 69B(1965)49-53.

17. Van Der Waerden Moderne Algebra Berlin 1937.

18. D. Welsh Matroid Theory London Math.Soc. Monograph 8, 1976.

19. H. Whitney On the Abstract Properties of Linear Dependence. Amer. J. Math. 57(1935) 509-533.

PART IV

DESIGNS

PARTITIONS OF X^2 AND OTHER TOPICS

(Notes from the talk of D.G. Higman)

1. EXAMPLES

a. Let $\alpha \subseteq X^2$ be a symmetric relation and ρ_α be the distance function in the graph (X,α). We set $\alpha^{(i)} = \{(x,y) \in X^2 : \rho_\alpha(x,y) = i\}$ and denote by $\mathfrak{O} = \{\alpha^{(0)}, \alpha^{(1)}, \alpha^{(2)}, \ldots\}$ the partition of X^2 induced by (X,α).

b. $X = A^n$, $\alpha = \{(x,y) \in X^2 : x,y$ differ in one coordinate$\}$. ρ_α is the Hamming metric, and $(X,\mathfrak{O})$ is called hypercubical association scheme or Hamming scheme.
In general, $|\mathfrak{O}|$ is called the rank of the scheme $(X,\mathfrak{O})$, $\mathfrak{O} \subseteq \mathcal{P}(X^2)$.

c. $X = A\{n\}$ (the set of all n-subsets of A), $\alpha = \{(x,y) : x \cap y = n-1\}$. $(X,\mathfrak{O})$ is called (hyper) tetrahedral scheme or Johnson scheme. Both schemes in b) and c) are of rank n.

d. $X = GF(q)$, $GF(q)^* = \langle w \rangle$, $s|q-1$, $H = \langle w^s \rangle$. Set $C_0 = \{0\}$, $C_1 = H$, $C_2 = Hw, \ldots,$ $\alpha_i = \{(x,y) \in X^2 : x-y \in C_i\}$. $(X,\mathfrak{O})$ is called cyclotomic scheme.

e. Group case. If the group G acts on X, set $\mathfrak{O}$ = totality of G-orbits on X^2.
Examples b),c),d) are of this type. In b) take $G = \Sigma_A \wr \Sigma_n$ (wreath product of two symmetric groups), in c) $G = \Sigma_A$, in d) $G = GF(q)\langle w^s \rangle$.

f. Flag configurations. X = set of flags in a finite projective plane, $\mathfrak{O} = \{1, \alpha, \beta, \alpha\beta, \beta\alpha, \alpha\beta\alpha = \beta\alpha\beta\}$, where $\alpha = \{((P,g),(P',g')) \mid P = P'\}$ and $\beta = \{((P,g),(P',g')) \mid g = g'\}$.

M. Aigner (ed.), Higher Combinatorics, 179-181.

2. STABILITY AND COHERENCE

For a partition $\mathcal{O}$ of X^2 set $\mathcal{O}^* = \mathcal{O} \wedge \{1, X^2-1\}$, where 1 is the diagonal. Hence $\mathcal{O} = \mathcal{O}^* \Leftrightarrow 1 = \bigcup E_i$, $E_i \in \mathcal{O}$. Further $\mathcal{O}^\cup = \{\alpha^\cup : \alpha \in \mathcal{O}\}$ where $\alpha^\cup = \{(y,x) : (x,y) \in \alpha\}$. Thus $\mathcal{O} = \mathcal{O}^\cup \Leftrightarrow \alpha = \alpha^\cup$ or $\alpha \wedge \alpha^\cup = 0$, for each $\alpha \in \mathcal{O}$.

For $\alpha, \beta \in \mathcal{O}$, $x, y \in X$ set
$$a_{\alpha\beta}(x,y) = |\{z \in X : (x,z) \in \alpha, (z,y) \in \beta\}| = |\alpha(x) \cap \beta^\cup(y)|,$$
and define the partition $\nabla\mathcal{O}$ by

$$(x,y) \sim (x_1,y_1) \bmod \nabla\mathcal{O} \Leftrightarrow a_{\alpha\beta}(x,y) = a_{\alpha\beta}(x_1,y_1), \text{ f.a. } \alpha,\beta \in \mathcal{O}.$$

We have

$$\mathcal{O} = \mathcal{O}^* \Rightarrow \nabla\mathcal{O} \le \mathcal{O}, \quad \mathcal{O}_1 \le \mathcal{O}_2 \Rightarrow \nabla\mathcal{O}_1 \le \nabla\mathcal{O}_2,$$

$$\mathcal{O} \le \nabla\mathcal{O} \Leftrightarrow \bigwedge_{\alpha,\beta,\gamma\in\mathcal{O}} a_{\alpha\beta}(x,y) =: a_{\alpha\beta\gamma} \text{ is independent of the choice of } (x,y) \in \gamma.$$

<u>Definition</u>. $\mathcal{O}$ is called ∇-stable if $\mathcal{O} \le \nabla\mathcal{O}$. $\mathcal{O}$ is coherent if $\mathcal{O} = \mathcal{O}^* = \mathcal{O}^\cup$ and $\mathcal{O}$ is ∇-stable.

Remark. Group induced $\underset{\not\Leftarrow}{\Rightarrow}$ coherent $\underset{\not\Leftarrow}{\Rightarrow}$ ∇-stable.

<u>Proposition</u>. There is a unique coarsest ∇-stable refinement of $\mathcal{O}$. There is a unique group induced coarsest refinement of $\mathcal{O}$.

3. ADJACENCY AND INTERSECTION ALGEBRAS

For a partition $\mathcal{O}$ on X^2 define the vector space $V(X,\mathcal{O}) = \{\varphi : X^2 \to \mathbb{C} : \varphi$ constant on the blocks of $\mathcal{O}\}$. The set $\{\phi_\alpha : \alpha \in \mathcal{O}\}$ where $\phi_\alpha(x,y) = \begin{cases} 1 & (x,y) \in \alpha \\ 0 & (x,y) \notin \alpha \end{cases}$ is a basis of $V(X,\mathcal{O})$.

<u>Proposition</u>. $\mathcal{O}$ is ∇-stable $\Leftrightarrow$ $V(X,\mathcal{O})$ is an algebra with the produkt $\phi_\alpha\phi_\beta(x,y) = a_{\alpha\beta}(x,y)$.

$V(X,\mathcal{O})$ endowed with this product is called the adjacency algebra.

Further, if $\mathcal{O}$ is coherent, then $V := V(X,\mathcal{O})$ is a semi-simple algebra containing the identity, called the intersection algebra; the numbers $a_{\alpha\beta\gamma}$ with $\phi_\alpha\phi_\beta = \sum_\gamma a_{\alpha\beta\gamma}\phi_\gamma$ are the intersection numbers of V.

If $\zeta(\varphi) = \text{trace}\,\varphi$ for each $\varphi \in V$, then $\zeta = \sum_{i=1}^{m} z_i\zeta_i$, where $\zeta_1,\ldots,\zeta_m$

are the irreducible characters of V. The z_i's are the multiplicities of $(X,\mathcal{O})$. Let $e_i = \zeta_i(1)$ for $1 \le i \le m$. Using the orthogonality relations $\sum_{\alpha\in\mathcal{O}} \zeta_i(\phi_\alpha)\zeta_j(\tilde{\phi}_\alpha) = \delta_{ij}\frac{e_i}{z_i}$, $\tilde{\phi}_\alpha = \frac{1}{|\alpha|}\phi_\alpha^t$ (valid when $1 \in \mathcal{O}$) and the character table of V we obtain necessary conditions for the existence of a coherent partition $(X,\mathcal{O})$.

4. ASSOCIATION SCHEMES

Let $(X,\mathcal{O})$ be coherent. We say $(X,\mathcal{O})$ is homogeneous if $1 \in \mathcal{O}$, $(X,\mathcal{O})$ is an association scheme if V is commutative, $(X,\mathcal{O})$ is a symmetric association scheme if $\alpha = \alpha^\cup$, $\alpha \in \mathcal{O}$, $(X,\mathcal{O})$ is a metric association scheme, if $\alpha \in \mathcal{O}$ is induced by a distance ρ. From 2. we deduce

metric ⇒ symmetric ⇒ association scheme ⇒ homogenous.

REFERENCES

1. Bose, R.C., Combinatorial properties of partially balanced designs and association schemes. Sankhya 25(1963), 109-136.

2. Cameron, P.J. - Van Lint, J.H., Graph Theory, Coding Theory and Block Designs. London Math. Soc. Lecture Note Ser. 19 (1975), Cambridge Univ. Press.

3. Delsarte, P., An algebraic approach to the association schemes of coding theory. Philips Research Reports Supplements 1973, No. 10.

4. Higman, D.G., Combinatorial considerations about permutation groups. Mathematical Institute, Oxford (1971).

5. Snapper, E., Group characters and integral matrices. J. Algebra 19 (1971), 520-535.

COMBINATORICS OF PARTIAL GEOMETRIES AND GENERALIZED QUADRANGLES

J.A. Thas

State University of Ghent

1. PARTIAL GEOMETRIES

1.1. Introduction

A finite partial geometry [7] is an incidence structure $S=(P,B,I)$, with a symmetric incidence relation satisfying the following axioms

(i) each point is incident with $1+t$ lines ($t\geq 1$) and two distinct points are incident with at most one line;

(ii) each line is incident with $1+s$ points ($s\geq 1$) and two distinct lines are incident with at most one point;

(iii) if x is a point and L is a line not incident with x, then there are exactly α ($\alpha\geq 1$) points $x_1, x_2, \ldots, x_\alpha$ and α lines $L_1, L_2, \ldots, L_\alpha$ such that $x I L_i I x_i I L$, $i=1,2,\ldots,\alpha$.

We say that the points x,y (resp. lines L,M) are collinear (resp. concurrent) if they are incident with at least one line (resp. point). If the points x,y (resp. lines L,M) of S are collinear (resp. concurrent), then we write $x\sim y$ (resp. $L\sim M$); otherwise we write $x\not\sim y$ (resp. $L\not\sim M$).

1.2. Theorem

Let $S=(P,B,I)$ *be a partial geometry with parameters* s,t,α . *If* $|P|=v$ *and* $|B|=b$, *then* $v=(s+1)(st+\alpha)/\alpha$ *and* $b=(t+1)(st+\alpha)/\alpha$.

Proof. Let L be a fixed line of S, and count on different ways the number of ordered pairs (x,M), with $x\not I L$, xIM and $L\sim M$. There results : $\alpha(v-s-1)=(s+1)ts$ or $v=(s+1)(st+\alpha)/\alpha$.

M. Aigner (ed.), Higher Combinatorics, 183-199.

Dually : $b=(t+1)(st+\alpha)/\alpha$.

Corollary. If S is a partial geometry with parameters s,t,α, then $\alpha \mid st(s+1)$ and $\alpha \mid st(t+1)$.

Remark. Let $S=(P,B,I)$ be a partial geometry with parameters s,t,α. If $E=\{\{x,y\} \parallel x,y\in P \text{ and } x\sim y\}$, then (P,E) is a strongly regular graph with parameters $v=(s+1)(st+\alpha)/\alpha$, $n_1=st+s$, $p^1_{11}=s-1+t(\alpha-1)$, $p^2_{11}=(t+1)\alpha$ [7].

1.3. Theorem (R.C. Bose [6] - D.G. Higman [20])

If $S=(P,B,I)$ *is a partial geometry with parameters* s,t,α, *then* $\alpha(s+t+1-\alpha) \mid st(s+1)(t+1)$

Proof. Let $P=\{x_1,x_2,\ldots,x_v\}$ and let $A=(a_{ij})$ be the vxv-matrix for which $a_{ij}=0$ if $i=j$ or $x_i \not\sim x_j$, and $a_{ij}=1$ if $i\neq j$ and $x_i\sim x_j$ (i.e. A is the adjacency matrix of the graph (P,E)). If $A^2=(c_{ij})$, then we have : (a) $c_{ii}=(t+1)s$ (b) if $i\neq j$ and $x_i\not\sim x_j$, then $c_{ij}=(t+1)\alpha$ (c) if $i\neq j$ and $x_i\sim x_j$, then $c_{ij}=s-1+t(\alpha-1)$. Consequently,

$$A^2-(s-t-\alpha-1)A-(t+1)(s-\alpha)I=(t+1)\alpha J$$

(I is the identity matrix of order v, J is the vxv-matrix with all entries equal to 1).

Evidently $(t+1)s$ is an eigenvalue of A, and J has eigenvalues $0,v$ with resp. multiplicities $v-1,1$. From $((t+1)s)^2-(s-t-\alpha-1)(t+1)s-(t+1)(s-\alpha)=(t+1)\alpha v$ follows that the eigenvalue $s(t+1)$ of A corresponds with the eigenvalue v of J, and consequently that $s(t+1)$ has multiplicity 1. The other eigenvalues of A are roots of the quadratic equation $x^2-(s-t-\alpha-1)x-(t+1)(s-\alpha)=0$. The multiplicities of these eigenvalues x_1,x_2 are denoted by m_1,m_2. We have $x_1=-t-1$, $x_2=s-\alpha$, and $1+m_1+m_2=v$, $s(t+1)-m_1(t+1)+m_2(s-\alpha)=\mathrm{tr}A=0$. Hence $m_2=(s+1)(t+1)st/\alpha(s+t+1-\alpha)$ and

$$m_1=s((s+1)t+\alpha)(s-\alpha+(t+1)\alpha)/\alpha(s+t+1-\alpha)$$

We conclude that $\alpha(s+t+1-\alpha) \mid st(s+1)(t+1)$.

1.4. The four classes of partial geometries

The class of partial geometries is divided into four (non disjoint) classes.

(a) The partial geometries with $\alpha=s+1$ or, dually, $\alpha=t+1$. The necessary conditions for the existence of such partial geometries become $s+1 \mid t(t+1)$ or, dually, $t+1 \mid s(s+1)$. A partial geometry S with $\alpha=s+1$ is the same as a 2-$(v,s+1,1)$ design.

(b) The partial geometries with $\alpha=s$, or, dually, $\alpha=t$. The necessary conditions are trivial in this case. A partial geometry S with $\alpha=t$ is the same as a net of order $s+1$ and deficiency $s-t+1$.

(c) The partial geometries with $\alpha=1$.
A partial geometry for which $\alpha=1$ is called a generalized quadrangle. Generalized quadrangles were introduced by J. Tits [54]. For a generalized quadrangle S with parameters s,t, we have $s+t \mid st(s+1)\times(t+1)$. This condition was also proved by W. Feit - G. Higman [18]. In the second part of this paper we shall study in detail these generalized quadrangles.

(d) The partial geometries with $1<\alpha<\min(s,t)$.

1.5. Partial geometries with $1<\alpha<\min(s,t)$

Introductory remark. Since all known partial geometries with $1<\alpha<\min(s,t)$ are constructed with the aid of maximal arcs in projective planes, we shall devote the following paragraph to some important results concerning maximal arcs.
Maximal arcs. In a finite projective plane of order q, any non-void set of l points may be described as a $\{l;n\}$-arc, where n ($n\neq 0$) is the greatest number of collinear points in the set. For given q and n ($n\neq 0$), l can never exceed $(n-1)(q+1)+1$, and an arc with that number of points will be called a maximal arc [3]. Equivalently, a maximal arc may be defined as a non-void set of points meeting every line in just n points or in none at all.

If K is a $\{nq-q+n;n\}$-arc (i.e. a maximal arc) of a projective plane π of order q, where $n\leqslant q$, then it is easy to prove that the set $K'=\{\text{lines } L \text{ of } \pi \parallel L\cap K=\phi\}$ is a $\{q(q-n+1)/n;q/n\}$-arc (i.e. a maximal arc) of the dual projective plane π^* of π. It follows immediately that, if the desarguesian projective plane PG(2,q) contains a $\{nq-q+n;n\}$-arc, $n\leqslant q$, then it also contains a $\{q(q-n+1)/n;q/n\}$-arc.

From the preceding there follows that a necessary condition for the existence of a maximal arc (as a proper subset of a given plane) is that n should be a factor of q. R.H.F. Denniston [17] has proved that the condition does suffice in the case of any desarguesian plane of order 2^h. His construction is the following : consider an irreducible homogeneous quadratic form F(X,Y) over $GF(2^h)$ and also a subgroup H, of order $n=2^m$ ($0\leqslant m\leqslant h$), of the additive group A of $GF(2^h)$; if we choose a system of non-homogeneous coordinates (X,Y) in the desarguesian plane $PG(2,2^h)$, then $K=\{(X,Y) \parallel F(X,Y)\in H\}$ is a $\{2^{h+m}-2^h+2^m;2^m\}$-arc of $PG(2,2^h)$. But the condition $n|q$ does not suffice in the case of a desarguesian plane of odd order. A. Cossu [14] has proved that there is no $\{21;3\}$-arc in PG(2,9). In J.A. Thas [47] appears the following more general result : in PG(2,q), $q=3^h$ and $h>1$, there are no $\{2q+3;3\}$-arcs and no $\{q(q-2)/3;q/3\}$-arcs. We conjecture that in PG(2,q), q odd, there are no $\{nq-q+n;n\}$-arcs with $n<q$ (i.e. the only maximal arcs of PG(2,q), q odd, are PG(2,q) and AG(2,q)).

Finally we remark that, apart from Denniston's construction, there is only one other construction (to our knowledge) of maximal arcs with $n>2$ (in the case $n=2$ the maximal arc is a complete oval, and many constructions of complete ovals are known). This construction, which is in J.A. Thas [46], is the following. Consider an ovoid O and a 1-spread W of $PG(3,2^m)$ $(m>0)$, such that each line of W has one and only one point in common with O (i.e. such that W belongs to the linear complex of lines defined by O). Let $PG(3,2^m)$ be embedded as a hyperplane R in $PG(4,2^m)=P$, and let x be a point of P-R. Call C the set of the points of P-R which are on a line px, with $p\in O$. Then C is a maximal arc, with parameters $l=2^{3m}-2^{2m}+2^m$ and $n=2^m$, of the projective plane π defined by the 1-spread W. The known ovoids O of $PG(3,2^m)$ are the elliptic quadrics and the Tits-ovoids (a Tits-ovoid is only defined in a $PG(3,2^{2s+1})$, with $s\geq 1$), and the known 1-spreads W of $PG(3,2^m)$ which belong to a linear complex of lines are the desarguesian spreads (i.e. the regular spreads) and the Lüneburg-spreads (a Lüneburg-spread is only defined in a $PG(3,2^{2s+1})$, with $s\geq 1$) [44]. So we have four possibilities for the choice of the pair (O,W). In such a way there arise two types of maximal arcs in desarguesian planes, and two types in Lüneburg-planes. If O is an elliptic quadric and W a regular spread, then it is possible to prove that with the pair (O,W) corresponds a Denniston-arc (with the notations of the preceding paragraph we have $GF(2^h)=GF(2^{2m})$, $H=GF(2^m)$).

The known partial geometries with $1<\alpha<\min(s,t)$. Let K be a $\{qn-q+n;n\}$-arc, $1<n<q$, of a projective plane π (not necessarily desarguesian) of order q. Define points of the partial geometry S as the points of π which are not contained in K. Lines of S are the lines of π which are incident with n points of K. The incidence is that of π. Now it is easy to prove that the configuration S so defined is a partial geometry with parameters ([45], [46], [57]) : $s=q-n$, $t=q-q/n$, $\alpha=q-q/n-n+1$.

In particular, suppose that there exists a $\{qn-q+n;n\}$-arc K, $1<n<q$, in the projective plane PG(2,q) over GF(q). Then a second partial geometry S' is defined as follows. Let PG(2,q) be embedded as a plane H in PG(3,q)=P. Define points of the partial geometry as the points of P-H. Lines of S' are the lines of P which are not contained in H and meet K (necessarily in a unique point). The incidence is that of P. Now it is not difficult to prove that S' is a partial geometry with parameters $s=q-1$, $t=qn-q+n-1$, $\alpha=n-1$ [46].

As the existence of the $\{qn-q+n;n\}$-arc K in PG(2,q) implies the existence of a $\{q(q-n+1)/n;q/n\}$-arc K' in PG(2,q), there follows immediately that there also exists a partial geometry with parameters $s=q-1$, $t=(q^2-qn+q-1)/n$, $\alpha=(q-n)/n$ [46].

We remarked that there exist $\{2^{m+h}-2^h+2^m;2^m\}$-arcs in $PG(2,2^h)$ $(0<m<h)$, and hence there exist partial geometries with parameters

(a) $s=2^h-2^m$, $t=2^h-2^{h-m}$, $\alpha=(2^m-1)(2^{h-m}-1)$ $(0<m<h)$, and

(b) $s=2^h-1$, $t=2^{h+m}-2^h+2^m-1$, $\alpha=2^m-1$ $(0<m<h)$.

For such partial geometries there never holds $s+1=\alpha$, $t+1=\alpha$, $s=\alpha$ or $t=\alpha$. A partial geometry (a) is a generalized quadrangle iff $h=2$ and $m=1$ (then $s=t=2$, $v=b=15$). A partial geometry (b) is a generalized quadrangle iff $m=1$ (then $s=2^h-1$, $t=2^h+1$, $v=2^{3h}$, $b=2^{2h}(2^h+2)$), iff K is a complete oval. These quadrangles corresponding with complete ovals were first discovered by Ahrens and Szekeres [1] and independently by M. Hall, Jr. [19] .

1.6. Literature

For more information about partial geometries we refer to R.C. Bose [6], [8], C. Bumiller [11], F. De Clerck [15], [52] , D.G. Higman [20], and J.A. Thas [45], [46], [52], [39] .

2. GENERALIZED QUADRANGLES

2.1. Regularity and antiregularity

Let $S=(P,B,I)$ be a generalized quadrangle with parameters s,t . The following definitions are given for points, with the dual concepts for lines being assumed given also. If $x\in P$, then the star of x is defined by $st(x)=\{y\in P \parallel x\sim y\}$. If $x,y\in P$, $x\neq y$, then the trace of $\{x,y\}$ is defined by $tr(x,y)=\{z\in P \parallel z\sim x, z\sim y\}=st(x)\cap st(y)$, and the span of $\{x,y\}$ by $sp(x,y)=\{u\in P \parallel u\sim z\ \forall z\in tr(x,y)\}=\{u\in P \parallel tr(x,y)\subset st(u)\}$. We have $|st(x)|=st+s+1$, and if $x\not\sim y$ then $|tr(x,y)|=t+1$, and $|sp(x,y)|\leq t+1$ (if $x\sim y$, then $sp(x,y)=tr(x,y)$ and $|sp(x,y)|=s+1$). A set of three pairwise noncollinear points of S is called a triad (of points). Let $T=\{x,y,z\}$ be a triad. A point $u\in P$ for which $u\sim x$, $u\sim y$, $u\sim z$ is called a center of T. A triad having a center is called centric. A pair (x,y), $x\not\sim y$, is said to be regular provided $|sp(x,y)|=t+1$. The pair (x,y), $x\not\sim y$, is antiregular provided $|st(z)\cap tr(x,y)|\leq 2$ for all $z\in P-\{x,y\}$. A point x is regular (resp. antiregular) provided (x,y) is regular (resp. antiregular) for all $y\in P$ with $x\not\sim y$.

The proofs of the following theorems are given in [24], [24], [33], [32], [48] and [25] respectively :

(i) *If* $1<s<t$, *then* (x,y), $x\not\sim y$, *is neither regular nor antiregular.*

(ii) *The pair* (x,y), $x\not\sim y$, *is regular iff each centric triad* (x,y,z) *has exactly* 1 *or* 1+t *centers. When* s=1 *this is iff each triad* (x,y,z) *is centric.*

(iii) *Let* $s\geq t$ *and let* (x,y) *be a pair of noncollinear points.*

There is no triad (x,y,z) *with a unique center iff* t=1 *or* s=t *and* (x,y) *is antiregular.*

(iv) *If a generalized quadrangle* S *with parameters* s,s *has even one antiregular pair of points, then* s *is odd.*

(v) *If a generalized quadrangle* S *with parameters* s,s *has a regular point* x *and a regular pair* (L_1, L_2) *of nonconcurrent lines for which* x *is incident with no line of* $tr(L_1, L_2)$ *, then* s *is even.*

(vi) *Let* x *be a coregular point of a generalized quadrangle* S *with parameters* s,s *(i.e. each line through* x *is regular). Then* x *is regular if and only if* s *is even.*

The following remarks were first made by C.T. Benson [5] (see also [21]). Let S=(P,B,I) be a generalized quadrangle with parameters s,s (s>1), and suppose that S contains a regular point x. If L∈B, then the set of points incident with L is denoted by L^*. Let $B_x = B_1 \cup B_2$, where B_1 is the set of sets L^* with xIL, and B_2 is the set of spans contained in st(x). *Then the incidence structure* $(st(x), B_x, \in)$ *is a* $2-(s^2+s+1, s+1, 1)$ *design, i.e. a projective plane of order* s .

Next let us suppose that S contains an antiregular point x. We choose a point y, where y~x and y≠x. The following notations are introduced : $P^* = \{z \in st(x) \parallel z \not\sim y\}$ and $B^* = B_1 \cup B_2$, where B_1 is the set of sets $L^* - \{x\}$ with xIL, y$\not$IL, and where $B_2 = \{tr(x,z) - \{y\} \parallel z \sim y, z \neq x\}$. *Then the incidence structure* $(P^*, B^*, \in)$ *is a* $2-(s^2, s, 1)$ *design, i.e. an affine plane of order* **s**.

2.2. The inequality of D.G. Higman

Theorem (D.G. Higman [20]). *If* S=(P,B,I) *is a generalized quadrangle with parameters* s *and* t, *where* s>1 *and* t>1, *then* $t \leq s^2$, *and dually* $s \leq t^2$.

Proof (Cameron [12]). Let x,y be two noncollinear points of S. If $V = \{z \in P \parallel z \not\sim x, z \not\sim y\}$, then $|V| = d = (s+1)(st+1)-2-2(t+1)s+(t+1)$. The elements of V are denoted by $z_1, z_2, \ldots, z_d$, and let $t_i = |\{u \in tr(x,y) \parallel u \sim z_i\}|$.

We count on two different ways the number of ordered pairs (z_i, u), where $z_i \in V$, $u \in tr(x,y)$, $u \sim z_i$. Then we obtain $\sum_i t_i = (t+1)(t-1)s$ (1). Next we count on two different ways the number of ordered triples (z_i, u, u'), where $z_i \in V$, $u \in tr(x,y)$, $u' \in tr(x,y)$, $u \neq u'$, $u \sim z_i$, $u' \sim z_i$. Then we obtain $\sum_i t_i(t_i - 1) = (t+1)t(t-1)$ (2). From (1) and (2) follows that $\sum_i t_i^2 = (t+1)(t-1)(s+t)$. As $d\sum_i t_i^2 - (\sum_i t_i)^2 \geq 0$, there holds $d(t+1)(t-1)(s+t) \geq (t+1)^2(t-1)^2 s^2$ or $d(s+t) \geq (t+1)(t-1)s^2$. Consequently $t(s-1)(s^2-t) \geq 0$ or $s^2 \geq t$.

Corollary (R.C. Bose [6]) . Let S be a generalized quadrangle with parameters s and t, where $s>1$ and $t>1$. Then $s^2=t$ iff $d\sum_i t_i^2-(\sum_i t_i)^2=0$ for any pair of noncollinear points (x,y), iff $t_i=(\sum_i t_i)/d$ for all $i\in\{1,2,\dots,d\}$ (and for any pair of noncollinear points (x,y)), iff each triad of points has a constant number of centers. This constant number of centers equals s+1. Let $s^2=t$ ($s>1$), and let (x,y,z) be a triad of points. Then we define the trace of (x,y,z) to be the set of the centers of (x,y,z). The span of (x,y,z) is the set $sp(x,y,z)=\{u\in P \parallel u\sim w\ \forall w\in tr(x,y,z)\}$. From the preceding follows that $|tr(x,y,z)|=s+1$ and $|sp(x,y,z)|\leq s+1$. If $|sp(x,y,z)|=s+1$, then we say that (x,y,z) is regular.

The proof of P.J. Cameron may be generalized to obtain the following theorem, which appeared to be fundamental for the combinatorial characterizations of the classical quadrangles.

Theorem (J.A. Thas [40]). *If for some pair* (x,y), $x\not\sim y$, *we have* $|sp(x,y)|\geq p+1$, *with* p *a positive integer, then* $s=1$ *or* $pt\leq s^2$. *Moreover, if* $p\neq t$ *then the number of centers of the triad* (x,y,z), *where* z *is collinear with no point of* sp(x,y), *is a constant iff* $pt=s^2$ (*this constant equals* (s+p)/p) .

2.3. Subquadrangles

The generalized quadrangle S'=(P',B',I') is called a subquadrangle of the generalized quadrangle S=(P,B,I) if and only if $P'\subset P$, $B'\subset B$ and $I'=I\cap((P'\times B')\cup(B'\times P'))$. If $S'\neq S$, then we say that S' is a proper subquadrangle of S. The following basic theorem first appeared in S.E. Payne [26], but the proof we shall give is due to J.A. Thas [50] .

Theorem (S.E. Payne - J.A. Thas). *If the generalized quadrangle* S'=(P',B',I') *with parameters* s',t' *is a subquadrangle of the generalized quadrangle* S=(P,B,I) *with parameters* s,t, *then* $s=s'$ *or* $s\geq s't'$ *and dually* $t=t'$ *or* $t\geq s't'$.

Proof. Suppose that S' is a proper subquadrangle of S. Then $P\neq P'$ and $B\neq B'$ (from P=P' (resp. B=B') follows that $(1+s)(1+st)=(1+s')(1+s't')$ (resp. $(1+t)(1+st)=(1+t')(1+s't')$), with $s\geq s'$ and $t\geq t'$, and so s=s' and t=t', i.e. S=S', a contradiction). Let xIL, $x\in P-P'$ and $L\in B'$. If xIM, $M\in B$ and $L\neq M$, then from the definition of generalized quadrangle follows immediately that M is incident with no point of P'. Further let $V=\{x\in P-P' \parallel \nexists L\in B' \text{ with } xIL\}$. Then $|V|=d=(1+s)(1+st)-(1+s')(1+s't')-(1+t')(1+s't')(s-s')$.

(a) t=t'. As $|V|=d\geq 0$ we have $(s-s')t(s-s't)\geq 0$. Consequently $s=s'$ or $s\geq s't$.

(b) $t>t'$. Let $V=\{x_1,x_2,\dots,x_d\}$ and let t_i be the number of points of P' which are collinear with x_i . We count on two different ways the number of ordered pairs (x_i,z), $x_i\in V$, $z\in P'$, $x_i\sim z$.

Then we obtain $\sum_i t_i=(1+s')(1+s't')(t-t')s$ (1). Next we count on two different ways the number of ordered triples (x_i, z, z'), $x_i \in V$, $z \in P'$, $z' \in P'$, $z \neq z'$, $x_i \sim z$, $x_i \sim z'$. Then we obtain $\sum_i t_i(t_i-1)= =(1+s')(1+s't')s'^2t'(t-t')$ (2). From (1) and (2) follows that $\sum_i t_i^2=(1+s')(1+s't')(t-t')(s+s'^2t')$. As $d\sum_i t_i^2-(\sum_i t_i)^2 \geqslant 0$, we obtain $(1+s')(1+s't')(t-t')(s-s')(s-s't')(st+s'^2t'^2) \geqslant 0$. Since $t>t'$ there results that $s=s'$ or $s \geqslant s't'$.

From (a) and (b) follows immediately that $s=s'$ or $s \geqslant s't'$. Dually, $t=t'$ or $t \geqslant s't'$.

<u>Corollary</u>. Suppose that $t>t'$. From (b) follows that $t_i=(\sum_i t_i)/d$ $\forall i \in \{1,2,\ldots,d\}$ if and only if $d\sum_i t_i^2-(\sum_i t_i)^2=0$, i.e. if and only if $s=s'$ or $s=s't'$. If $s=s'$, then $t_i=1+st'\ \forall i$; if $s=s't'$, then $t_i=1+s'\ \forall i$. This can easily be dualized.

The next results are easy consequences of the inequality of D.G. Higman and the preceding theorem, although they first appeared in J.A. Thas [51].

<u>Theorem (J.A. Thas - S.E. Payne)</u>. *Let* S' *be a proper subquadrangle of* S *with* S *having parameters* s,t *and* S' *having parameters* s,t'.

(i) $t \geqslant s$. *If* $t=s$, *then* $t'=1$.

(ii) *If* $s>1$, *then* $t' \leqslant s$. *If* $t'=s \geqslant 2$, *then* $t=s^2$.

(iii) *If* $s=1$, *then* $1 \leqslant t' < t$ *is the only restriction on* t'.

(iv) *If* $s>1$ *and* $t'>1$, *then* $\sqrt{s} \leqslant t' \leqslant s$ *and* $s^{3/2} \leqslant t \leqslant s^2$.

(v) *If* $t=s^{3/2}>1$ *and* $t'>1$, *then* $t'=\sqrt{s}$.

(vi) *Let* S' *have a proper subquadrangle* S" *with parameters* s,t'', $s>1$. *Then* $t''=1$, $t'=s$, *and* $t=s^2$.

Another theorem which appeared to be very useful for several characterization theorems is the following [51].

<u>Theorem</u>. *Let* S'=(P',B',I') *be a substructure of the generalized quadrangle* S=(P,B,I) *with parameters* s *and* t, ($s>1$), *for which the following conditions are satisfied :*

(i) *if* $x,y \in P'$ ($x \neq y$) *and* xILIy, *then* $L \in B'$;

(ii) *each element of* B' *is incident with* s+1 *elements of* P'.

Then there are three possibilities :

(a) *the elements of* B' *are lines which are incident with a distinguished point of* P, *and* P' *consists of the points of* P *which are incident with these lines;*

(b) $B'=\emptyset$ *and* P' *is a set of pairwise noncollinear points of* P;

(c) S' *is a subquadrangle with parameters* s *and* t'.

2.4. All known quadrangles

<u>The classical generalized quadrangles</u>. (a) We consider a non-singular hyperquadric Q of index 2 of the projective space PG(d,q), with d=3,4 or 5. Then the points of Q together with the lines of Q (which are the subspaces of maximal dimension on Q) form a

generalized quadrangle $Q(d,q)$ with parameters [16]

$s=q$, $t=1$, $v=(q+1)^2$, $b=2(q+1)$, when $d=3$;

$s=t=q$, $v=b=(q+1)(q^2+1)$, when $d=4$;

$s=q$, $t=q^2$, $v=(q+1)(q^3+1)$, $b=(q^2+1)(q^3+1)$, when $d=5$.

We remark that the structure of $Q(3,q)$ is trivial.

(b) Let H be a non-singular hermitian primal of the projective space $PG(d,q^2)$, $d=3$ or 4. Then the points of H together with the lines on H form a generalized quadrangle $H(d,q^2)$ with parameters [16].

$s=q^2$, $t=q$, $v=(q^2+1)(q^3+1)$, $b=(q+1)(q^3+1)$, when $d=3$;

$s=q^2$, $t=q^3$, $v=(q^2+1)(q^5+1)$, $b=(q^3+1)(q^5+1)$, when $d=4$.

(c) The points of $PG(3,q)$, together with the totally isotropic lines with respect to a symplectic polarity, form a generalized quadrangle $W(q)$ with parameters $s=t=q$, $v=b=(q+1)(q^2+1)$ [16].

All these generalized quadrangles (all of which are associated with classical simple groups) are due to J. Tits.

Isomorphisms between the classical generalized quadrangles. It was proved that the generalized quadrangle $Q(4,q)$ is isomorphic to the dual of $W(q)$ [5], and that $Q(5,q)$ is isomorphic to the dual of $H(3,q^2)$ [33]. Moreover $W(q)$ is selfdual if and only if q is even [44].

Regular points and lines of the classical models. Each line and each point of $Q(3,q)$ are regular. Moreover each point of $Q(3,q)$ is antiregular.

Each line of $Q(4,q)$ is regular. If q is even then each point of $Q(4,q)$ is regular; if q is odd then each point of $Q(4,q)$ is antiregular.

Each line of $Q(5,q)$ is regular. Moreover each triad of points is regular.

$H(4,q^2)$ contains no regular point and no regular line. But for any pair (x,y) of noncollinear points we have $|sp(x,y)|=q+1$. Let z be a point which is collinear with no point of $sp(x,y)$. Since $(|sp(x,y)|-1)t=s^2$, the number of centers of the triad (x,y,z) equals $(s+q)/q=q+1$.

Some important subquadrangles for which s=s' or t=t'. $Q(4,q)$ has subquadrangles isomorphic to $Q(3,q)$ (here $s=t$ and $t'=1$). $Q(5,q)$ has subquadrangles isomorphic to $Q(4,q)$ (here $s=t'$, $s^2=t$). $H(4,q^2)$ has subquadrangles isomorphic to $H(3,q^2)$ (here $t=s^{3/2}$, $t'=\sqrt{s}$). $H(3,q^2)$ has subquadrangles isomorphic to $W(q)$ (dual of (b)).

All known non classical quadrangles. (a) Let $P=\{x_{ij} \| i,j=1,2,\dots,s+1\}$ and $B=\{L_1,\dots,L_{s+1},M_1,\dots,M_{s+1}\}$, with $s\geq 1$. Incidence is defined as follows : $x_{ij} I L_k \Leftrightarrow i=k$, $x_{ij} I M_k \Leftrightarrow j=k$. Then $S=(P,B,I)$ is a generalized quadrangle with parameters $s=s$, $t=1$, $v=(s+1)^2$, $b=2(s+1)$.
Regular points and regular lines. Each point and each line of S are regular. Moreover each point of S is antiregular.
Subquadrangles. It is an easy exercise to prove that S has a subquadrangle with parameters s' and 1, for any integer s' with $1\leq s'\leq s$.

(b) (J. Tits [16]). Let d=2 (resp. d=3) and consider an oval O (resp. an ovoid O) of PG(d,q). Let PG(d,q) be embedded as a hyperplane H in PG(d+1,q)=P. Define points as (i) the points of P-H, (ii) the hyperplanes X of P for which $|X\cap O|=1$, and (iii) one new symbol x. Lines are (a) the lines of P which are not contained in H and meet O (necessarily in a unique point), and (b) the points of O. Incidence is defined as follows : Points of type (i) are incident only with lines of type (a); here the incidence is that of P. A point X of type (ii) is incident with all lines $\subset X$ of type (a) and with precisely one line of type (b), namely the one represented by the unique point of O in X. Finally, the unique point x of type (iii) is incident with no line of type (a) and all lines of type (b). The incidence structure so defined is a generalized quadrangle with parameters

$s=t=q$, $v=b=(q+1)(q^2+1)$, when d=2;
$s=q$, $t=q^2$, $v=(q+1)(q^3+1)$, $b=(q^2+1)(q^3+1)$, when d=3.

If d=2 the quadrangle is denoted by $T_2(O)$; if d=3 the quadrangle is denoted by $T_3(O)$. If no confusion is possible these quadrangles are also denoted by T(O).

Further information about $T_2(O)$. $T_2(O)$ is isomorphic to Q(4,q) if and only if O is an irreducible conic of PG(2,q). By the celebrated theorem of B. Segre [34] each oval O of PG(2,q), q odd, is an irreducible conic. So for q odd $T_2(O)$ is always isomorphic to Q(4,q).

Let us suppose that q is even. Then not every oval is an irreducible conic [34] . So in this case there are quadrangles $T_2(O)$ which are not isomorphic to Q(4,q). We remark that the point x of type (iii) is regular, and that all lines of type (b) are regular. Consequently $T_2(O)$ has always subquadrangles with parameters s'=q, t'=1, and also with parameters s'=1, t'=q.

Further information about $T_3(O)$. $T_3(O)$ is isomorphic to Q(5,q) if and only if O is an elliptic quadric of PG(3,q). By the celebrated theorem of A. Barlotti [2] each ovoid O of PG(3,q), q odd, is an elliptic quadric. So for q odd $T_3(O)$ is always isomorphic to Q(5,q).

Let us suppose that q is even. If q is an odd power of two, then there are ovoids which are not elliptic quadrics [55] . So in the even case there are quadrangles $T_3(O)$ which are not isomorphic to Q(5,q). We remark that all lines of type (b) are regular, and that all triads (x,y,z), with x the unique point of type (iii), are regular. Also it is easy to show that $T_3(O)$ has always subquadrangles with parameters s'=t'=q.

(c) (S.E. Payne - R.W. Ahrens and G. Szekeres - M. Hall, Jr.) The following important construction method is due to S.E. Payne ([27] and [28]).

Let x be a regular point of the generalized quadrangle S=(P,B,I) with parameters s,s . P' is defined to be the set P-st(x). The elements of B' are of two types : the elements of type (a) are the lines of B which are not incident with x; the elements of type (b) are the spans sp(x,y), with $y \not\sim x$. The incidence relation I' is defined as follows : if $y \in P'$ and L is a line of type (a) of B', then yI'L iff yIL; if $y \in P'$ and L is a line of type (b) of B', then yI'L iff $y \in L$. Now it is not hard to prove that S'=(P',B',I') is a generalized quadrangle with parameters s'=s-1, t'=s+1, $v'=s^3$, $b'=s^2(s+2)$. We say that S' is obtained by expanding the quadrangle S about the regular point x.

Let S=W(q), q odd. Then every point is regular and every line antiregular. Expanding about any point yields the only known quadrangle with parameters q-1 and q+1, first discovered by Ahrens and Szekeres [1] . Next let $S=T_2(O)$, with q even. The unique point of type (iii) is regular, and all lines through it are regular. Expanding about this regular point yields a quadrangle first discovered by Ahrens and Szekeres and independently by Hall (see 1.5.). Expanding in a dual fashion about one of the regular lines sometimes yields new quadrangles with parameters q+1 and q-1 [28] .

Remark. For the parameters of any of the known quadrangles, we have :

s=1, $t \geq 1$ and dually t=1, $s \geq 1$, or
s=t=q, with q a prime power, or
s=q, $t=q^2$ and dually $s=q^2$, t=q, with q a prime power, or
$s=q^2$, $t=q^3$ and dually $s=q^3$, $t=q^2$, with q a prime power, or
s=q-1, t=q+1 and dually s=q+1, t=q-1, with q a prime power.

2.5. Generalized quadrangles with a small number of points and lines

Now we consider which generalized quadrangles S *with parameters* s *and* t, s≤t *and* s *small, can exist and/or are known.*

(a) s=1. Then for any t≥1 there is a unique example.

(b) s=2. By 1.3. and 2.2. either t=2 or t=4. In either case there is a unique quadrangle. The t=2 case was probably first considered by C.T. Benson [5]. The t=4 case was handled independently at least three times, by J.J. Seidel [36], E. Shult [37], and J.A. Thas [42].

(c) s=3. Here t=3,5,6, or 9. S.E. Payne has proved that for t=3 there is an example unique up to duality [30]. For t=5 there is a known example in the infinite class with parameters q-1, q+1, with q any prime power. Uniqueness is undecided as far as we know. For t=6 nothing is known. P.J. Cameron was the first to prove that Q(5,3) is the unique quadrangle with parameters 3 and 9 ([13],[32]).

(d) s=4. Here t=4,6,8,11,12 or 16. Nothing is known about 11 and 12. Recently S.E. Payne proved that there is a unique quadrangle with parameters 4,4 [31]. Unique examples are known for t=6,8 and 16, but the uniqueness question is far from settled.

2.6. Characterization theorems

Now we list the most important combinatorial characterizations of the known generalized quadrangles. We remark that the notions span, regular, antiregular, and subquadrangle play a central role in these theorems.

(i) (F. Buekenhout - C. Lefèvre [9]). If a pointset of the projective space PG(d,q), together with a lineset of PG(d,q) form a generalized quadrangle S, then S is a classical generalized quadrangle.

(ii) (C.T. Benson [5]). If s=t (>1) then S is isomorphic to W(s) if and only if all its points are regular.

(iii) (J.A. Thas [41]). If S is a generalized quadrangle for which $|sp(x,y)|\geq s+1>2\ \forall x,y\ (x\not\sim y)$, then S is isomorphic to W(s).

(iv) (F. Mazzocca - S.E. Payne - J.A. Thas [32]). Let S be a generalized quadrangle with parameters s,s having an antiregular point x. If there is a point $y\sim x$, $y\neq x$, for which the associated affine plane (see 2.1.) is desarguesian, then S is isomorphic to Q(4,s).

(v) (G. Tallini [38]). Let S=(P,B,I) be a generalized quadrangle for which each point is regular. If L=(P,{spans}, natural incidence relation), then a non-trivial subspace of L is a pointset $T \neq P$ which contains at least three points which do not belong to a same span, and for which $\{x,y\} \subset T$ implies that $sp(x,y) \subset T$.

A generalized quadrangle $S=(P,B,I)$ with $s>t$, $s>1$ and $t>1$ is isomorphic to an $H(3,q^2)$ iff the following conditions are satisfied

(a) each point of S is regular;

(b) $sp(x,y) \cup sp(x',y') \subset$ non-trivial subspace of L $\Rightarrow$ $tr(x,y) \cup \cup tr(x',y') \subset$ non-trivial subspace of L.

(vi) (J.A. Thas [43] and [49]). If the generalized quadrangle S with parameters $s=q$ and $t=q^2$ $(q>1)$ possesses a point x for which each triad (x,y,z) is regular, then S is isomorphic to a $T_3(O)$ (so if q is odd, then S is isomorphic to $Q(5,q)$). If q is even and if moreover each line is regular, then S is isomorphic to $Q(5,q)$.

(vii) (J.A. Thas [43]). If $s^2=t$ (>1), then S is isomorphic to $Q(5,s)$ if and only if every triad is regular.

(viii) (J.A.Thas [49]). Suppose that the generalized quadrangle $S=(P,B,I)$ with parameters s,t $(s>1)$ contains a point x such that every centric triad of lines, with a center incident with x, is contained in a proper subquadrangle S' with parameters s,t'. Then S has parameters s,s^2 and is isomorphic to a quadrangle $T_3(O)$ (so if s is odd, then S is isomorphic to $Q(5,s)$).

(ix) (J.A. Thas [41]). If S is a generalized quadrangle with parameters s,t $(s>1)$ for which every centric triad of lines is contained in a proper subquadrangle S' with parameters s,t', then S is isomorphic to $Q(5,s)$.

(x) (J.A. Thas [49]). First of all we introduce the following axiom which we denote by (D) : if L_1, L_2, M_1, M_2, U_1, U_2, N_1 are lines for which $L_1 \sim L_2$, $M_1 \sim M_2$, $L_1 \sim U_1$, $M_1 \sim U_1$, $L_2 \sim U_2$, $M_2 \sim U_2$, $U_1 \sim U_2$, $L_1 \sim N_1$, $M_1 \sim N_1$, then there exists at least one line N_2 for which $L_2 \sim N_2$, $M_2 \sim N_2$, $N_1 \sim N_2$.

Suppose that the generalized quadrangle S with parameters s,t, where $2<s<t$, contains a point x, such that each line incident with x is regular and such that (D) is satisfied whenever L_1 or L_2 contains x. Then S has parameters s,s^2 and is isomorphic to a $T_3(O)$ (so for s odd S is isomorphic to $Q(5,s)$).

(xi) (J.A. Thas [49]). If S is a generalized quadrangle with parameters s,t, where $2<s<t$, for which each line is regular and for which (D) is satisfied, then $s^2=t$ and S is isomorphic to $Q(5,s)$.

(xii) (J.A. Thas [40]). If $s^3=t^2$ (>1) then S is isomorphic to

H(4,s) if and only if $|sp(x,y)| \geq \sqrt{s}+1 \; \forall x,y \; (x \not\sim y)$.

(xiii) (J.A. Thas [41]). Let S be a generalized quadrangle with parameters s,t for which the following conditions are satisfied

(a) for each sp(x,y), $x \not\sim y$, there holds that each centric triad (x,y,z), where z is collinear with no point of sp(x,y), has at least two centers

(b) there exist points x and y, $x \not\sim y$, for which $t+1 > |sp(x,y)| > 2$.
Then S is isomorphic to H(4,s).

(xiv) (J.A. Thas [41]). If S is a generalized quadrangle for which $|sp(x,y)| \geq s^2/t+1 \; \forall x,y, \; x \not\sim y$, then one of the following occurs :

(a) s=1 (b) $t=s^2$ (c) S is isomorphic to W(s) (d) S is isomorphic to H(4,s).

(xv) (J.A.Thas [41]). Suppose that for the generalized quadrangle S the following condition is satisfied : for any sp(x,y), $x \not\sim y$, the number of centers of the triad (x,y,z), where z is collinear with no point of sp(x,y), is a constant. Then one of the following occurs : (a) all points of S are regular (b) $s^2=t$ (c) S is isomorphic to H(4,s).

(xvi) (J.A. Thas [41]). Let S be a generalized quadrangle with parameters s,t (s>1). If any set $\{x\} \cup tr(y,z)$ $(y \not\sim z)$, where x is collinear with no point of sp(y,z), is contained in a proper subquadrangle with parameters s,t', then S is isomorphic to W(s), Q(5,s) or H(4,s).

(xvii) (J.A. Thas [41]). Let S be a generalized quadrangle with parameters s,t, for which not every point is regular. If each set $\{x\} \cup tr(y,z)$ $(y \not\sim z)$, where x is collinear with no point of sp(y,z) and collinear with at least one point of tr(y,z), is contained in a proper subquadrangle with parameters s,t', then S is isomorphic to Q(5,s), H(4,s) or to Q(4,s) with s odd.

Remark. Other combinatorial characterizations which emphasize the connection between circle geometries (inversive, Minkowski and Laguerre planes) and generalized quadrangles can be found in [48] and [32] .

REFERENCES

1. R. Ahrens and G. Szekeres, On a combinatorial generalization of 27 lines associated with a cubic surface, J. Austral. Math. Soc. X(1969), 485-492 .

2. A. Barlotti, Un estensione del teorema di Segre-Kustaanheimo, Boll.Un. Mat. Ital.,(3) 10 (1955), 498-506.

3. A. Barlotti, Sui {k;n}-archi di un piano lineare finito, Boll. Un. Mat. Ital., (3) 11 (1956), 553-556.
4. A. Barlotti, Some topics in finite geometrical structures, Inst. of Stat. Mimeo Series No. 439, 1965
5. C.T. Benson, On the structure of generalized quadrangles, J. Algebra, 15 (1970), 443-454.
6. R.C. Bose, Graphs and designs, C.I.M.E., II ciclo, Bressanone 1972, 1-104.
7. R.C. Bose, Strongly regular graphs, partial geometries, and partially balanced designs, Pacific J. Math., 13 (1963), 389-419.
8. R.C. Bose, Geometric and pseudo-geometric graphs (q^2+1, $q+1,1$), J. Geom., 2 (1972), 75-93
9. F. Buekenhout and C. Lefèvre, Generalized quadrangles in projective spaces, Arch. Math. 25 (1974), 540-552.
10. F. Buekenhout, Une caractérization des espaces affins basée sur la notion de droite, Math. Z., 111 (1969), 367-371.
11. C. Bumiller Jr., Partial geometries and rank three groups, Ph. D. Dissertation, Yale University (1975).
12. P.J. Cameron, Partial quadrangles, Quart. J. Math. Oxford, (3) 25 (1974), 1-13.
13. P.J. Cameron, Private communication.
14. A. Cossu, Su alcune proprietà dei {k;n}-archi di un piano proiettivo sopra un corpo finito, Rend. Mat. e Appl. 20 (1961), 271-277.
15. F. De Clerck, Partial geometries, Ph. D. Dissertation, University of Ghent.
16. P. Dembowski, Finite geometries, Springer Verlag, 1968.
17. R.H.F. Denniston, Some maximal arcs in finite projective planes, J. Comb. Theory 6(1969) 317-319.
18. W. Feit and G. Higman, The nonexistence of certain generalized polygons, J. Algebra, 1 (1964), 114-131.
19. M. Hall, Jr., Affine generalized quadrilaterals, Studies in Pure Mathematics, ed. by L. Mirsky, Academic Press, (1971), 113-116.
20. D.G. Higman, Partial geometries, generalized quadrangles, and strongly regular graphs, in Atti Convegno di Geometria e sue Applicazioni, Perugia, 1971.
21. F. Mazzocca, Sistemi grafici rigati di seconda specie, Relazione N. 28, Istit. di Math. dell'Univ. di Napoli, 22 pages.
22. F. Mazzocca, Caratterizzazione dei sistemi rigati isomorfi ad una quadrica ellittica dello $S_{5,q}$, con q dispari, Rend. Accad. Naz. Lincei, 57 (1974), 360-368.
23. D. Olanda, Sistemi rigati immersi in uno spazio proiettivo, Relazione n. 26, Ist. Mat. Univ. Napoli, (1973) 1-21.
24. S.E. Payne, Nonisomorphic generalized quadrangles, J. Algebra, 18 (1971), 201-212.

25. S.E. Payne, Generalized quadrangles of even order, J.Algebra, 31 (1974), 367-391.
26. S.E. Payne, A restriction on the parameters of a subquadrangle, Bull. Amer. Math. Soc., 79 (1973), 747-748.
27. S.E. Payne, The equivalence of certain generalized quadrangles, J. Comb. Theory, 10 (1971), 284-289.
28. S.E. Payne, Quadrangles of order (s-1,s+1), J. Alg., 22 (1972), 97-119.
29. S.E. Payne, Finite generalized quadrangles : a survey, Proc. Int. Conf. Proj. Planes, Wash. State Univ. Press, (1973), 219-261.
30. S.E. Payne, All generalized quadrangles of order 3 are known, J. Comb. Theory, 18 (1975), 203-206.
31. S.E. Payne, Generalized quadrangles of order 4, to appear.
32. S.E. Payne and J.A. Thas, Generalized quadrangles with symmetry, Simon Stevin, 49 (1975-1976), 3-32 and 81-103.
33. S.E. Payne and J.A. Thas, Classical finite generalized quadrangles : A combinatorial study, to appear.
34. B. Segre, Lectures on modern geometry, Ed. Cremonese, Roma, 1961.
35. B. Segre, Ovals in a finite projective plane, Canad. J. Math., 7 (1955), 414-416.
36. J.J. Seidel, Strongly regular graphs with (-1,1,0) adjacency matrix having eigenvalue 3, Lin. Algebra and Appl., 1 (1968), 281-298.
37. E. Shult, Characterizations of certain classes of graphs, J. Comb. Theory (B), 13 (1972), 142-167.
38. G. Tallini, Strutture di incidenza dodate di polarità, Rend. Sem. Mat. Milano, 41 (1971), 41 pp.
39. J.A. Thas, On polarities of symmetric partial geometries, Arch. Math., 25 (1974), 394-399.
40. J.A. Thas, On generalized quadrangles with parameters $s=q^2$ and $t=q^3$, Geometriae Dedicata, 5 (1976), 485-496.
41. J.A. Thas, Combinatorial characterizations of the classical generalized quadrangles, Geometriae Dedicata, to appear.
42. J.A. Thas, On 4-gonal configurations with parameters $r=q^2+1$ and $k=q+1$, part I, Geometriae Dedicata, 3 (1974), 365-375.
43. J.A. Thas, 4-gonal configurations with parameters $r=q^2+1$ and $k=q+1$, part II, Geometriae Dedicata, 4 (1975), 51-59.
44. J.A. Thas, Ovoidal translation planes, Arch. Math. 23 (1972), 110-112.
45. J.A. Thas, Construction of partial geometries, Simon Stevin 46 (1973), 95-98.
46. J.A. Thas, Construction of maximal arcs and partial geometries, Geometriae Dedicata, 3 (1974), 61-64.
47. J.A. Thas, Some results concerning $\{(q+1)(n-1);n\}$-arcs and $\{(q+1)(n-1)+1;n\}$-arcs in finite projective planes of order q, J. Comb. Theory, 19 (1975), 228-232.
48. J.A. Thas, 4-gonal configurations, in Finite Geometrical

Structures and their Applications, C.I.M.E., II ciclo, Bressanone (1972), 249-263.

49. J.A. Thas, Combinatorial characterizations of generalized quadrangles with parameters s=q and $t=q^2$, to appear.

50. J.A. Thas, A remark concerning the restriction on the parameters of a 4-gonal subconfiguration, Simon Stevin, 48 (1974-75), 65-68.

51. J.A. Thas, 4-gonal subconfigurations of a given 4-gonal configuration, Rend. Accad. Naz. Lincei, 53 (1972), 520-530.

52. J.A. Thas and F. De Clerck, Some applications of the fundamental characterization theorem of R.C. Bose to partial geometries, Rend. Accad. Naz. Lincei, to appear.

53. J.A. Thas and P. De Winne, Generalized quadrangles in projective spaces, J. of Geometry, to appear.

54. J. Tits, Sur la trialité et certains groupes qui s'en déduisent, Publ. Math. I.H.E.S., Paris 2 (1959), 14-60.

55. J. Tits, Ovoïdes et groupes de Suzuki, Arch. Math. 13 (1962), 187-198.

56. J. Tits, Buildings and BN-pairs of spherical type, Springer-Verlag Lecture Notes #386, Berlin, 1974

57. W.D. Wallis, Configurations arising from maximal arcs, J. Comb. Theory A, 15 (1973), 115-119.

EDGE DECOMPOSITIONS OF COLORED GRAPHS

(Notes from the talk of R.M. Wilson)

Let $\mathcal{D}$ be a class of directed loopless graphs whose edges have been colored with c colors. By $\Delta_v(\lambda_1,\ldots,\lambda_c)$ we denote the digraph with v vertices any two of which are joined by λ_i edges (in both directions) of color i.

PROBLEM. Given $\mathcal{D}$ and $c,\lambda_1,\ldots,\lambda_c,v$, when is it possible to find a $\mathcal{D}$-decomposition of $\Delta_v(\lambda_1,\ldots,\lambda_c)$, i.e. a set of subgraphs of the large group Δ_v, each isomorphic to some member of $\mathcal{D}$ such that each edge of Δ_v occurs in exactly one of the subgraphs.

EXAMPLES

1. $c=1$, $\lambda_1=\lambda$; $\mathcal{D}=\{K_k\}$, i.e. two directed edges for each undirected edge. The decompositions correspond precisely to balanced incomplete block designs with parameters v,k,λ.

2. $c=\lambda=1$; $\mathcal{D}=\{$ 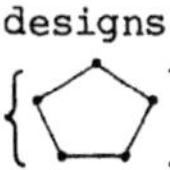 $\}$. The decompositions are called balanced circuit designs.

3. $c=\lambda=1$; $\mathcal{D}=\{$transitive tournament$\}$. We obtain permutation designs.

4. $c=1$, $\lambda_1=\lambda$; $\mathcal{D}=\{$  $\}$. This gives directed triple systems.

5. $c=2$, $\lambda_1=2$, $\lambda_2=1$; $\mathcal{D}=\{$ 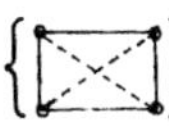 $\}$. We obtain arrangements for consecutive Whist matches. A Whist-factorization is a Whist-decomposition where the matches can be simultaneously arranged

M. Aigner (ed.), Higher Combinatorics, 201-202.

in rounds.

Example: $v = 8$, take $\mathbb{Z}_7 \cup \{\infty\}$, 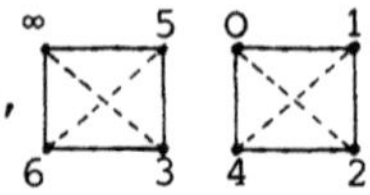

as initial round and develop modulo 7.

A necessary condition for the existence of Whist-factorizations is of course $v \equiv 0 \bmod 4$, and this is known to be sufficient except possibly for $v = 132, 152, 264$ (Hanani).

6. $c = 3$, $\lambda_1 = \lambda_2 = \lambda_3 = 1$. $\mathcal{D} = \{$ 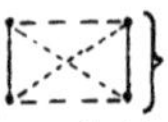 $\}$. These arrangements are called Triple Whist tournaments, and the corresponding factorizations exist for all but finitely many $v \equiv 0 \bmod 4$.

7. $c = 2$, $\lambda_1 = \lambda_2 = 1$, $\mathcal{D} = \{$ 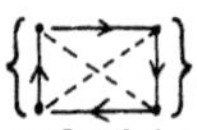 $\}$. We obtain directed Whist tournaments. Given a directed Whist decomposition define $x \circ x = x$, $x \circ y$ = person on x's left. This yields a quasigroup satisfying the identities

$$\begin{cases} x \circ x \\ (x \circ y) \circ (y \circ x) = y \end{cases},$$

and quasigroups satisfying these identities are combinatorially equivalent to the existence of directed Whist decompositions. The smallest value of v for which they exist is $v = 40$.

8. $c = 4$, $\lambda_1 = \lambda_2 = \lambda_3 = \lambda_4 = 1$, $\mathcal{D} = \{$ 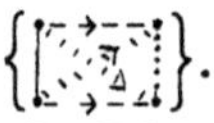 $\}$. Given a decomposition of $\Delta_v(1,1,1,1)$, define $x \circ x = x$, $x \circ y$ = vertex to the left of x. Such quasigroups are equivalent to selforthogonal Latin squares which are known to exist for all $v \neq 2, 3, 6$.

There followed a short discussion of the necessary conditions for the existence of a $\mathcal{D}$-decomposition Δ_v and remarks as to sufficiency proofs for $v \geq v_0$.

REFERENCES

1. Hanani, H., Balanced incomplete block designs and related designs. Discrete Math. 11 (1975), 255-369.

2. Wilson, R.M., An Existence Theory for Pairwise Balanced Designs I, II. J. Combinatorial Theory Ser.A 13 (1972), 220-245, 246-273. III. J.C.T.Ser.A 18 (1975), 71-79.

PART V

GROUPS AND CODING THEORY

PLANES AND BIPLANES

E. F. Assmus, Jr., Joseph A. Mezzaroba, and
Chester J. Salwach
Department of Mathematics, Lehigh University,
Bethlehem, Pennsylvania

It is possible to produce self-orthogonal codes from the incidence matrices of symmetric (v,k,λ)-designs. We describe four such methods here, where M is the incidence matrix of the design. Method 3, however, is the one which we exploit in the sequel.

Method 1: If $k \equiv \lambda \equiv 0 \pmod p$, then the row space of M over F_p is clearly a self-orthogonal code.

Method 2: If $p \mid (k-\lambda)$ where $p \nmid k$, let G be the $v \times (v+1)$ matrix whose first column is a column of $\sqrt{-k}$'s and whose last v columns are those of M, and let $F = F_p$ if $\left(\frac{-k}{p}\right) = 1$ and F_{p^2} otherwise. Then A, the rowspace of G over F, is a self-orthogonal code. Moreover, if $p^2 \nmid (k-\lambda)$, then $A = A^{\perp}$ and so A is a $(v+1, \frac{v+1}{2})$ self-dual code over F.

Proof: The first assertion is clearly true. Since A is self-orthogonal, $Rk_F(G) \leq \frac{v+1}{2}$. A standard calculation yields that $\det(M) = k(k-\lambda)^{\frac{v-1}{2}}$, and by the theory of elementary divisors there exist unimodular matrices P and Q such that PMQ is a diagonal matrix with $d_i \mid d_{i+1}$, where $1 \leq i \leq v-1$. Thus $Rk_F(G) = Rk_F(M) = Rk_F(PMQ) \geq v - \frac{v-1}{2} = \frac{v+1}{2}$ when $p^2 \nmid (k-\lambda)$; hence $Rk_F(G) = \frac{v+1}{2}$ and A is self-dual in this case.

M. Aigner (ed.), Higher Combinatorics, 205-212.

Method 3: If $p \mid \lambda$ and $k \equiv -1 \pmod{p}$, let G be the $v \times 2v$ matrix whose first v columns constitute the identity matrix and whose last v columns are those of M. Then the row space of G over F_p is a $(2v, v)$ self-dual code.

Method 4: If $p = 2$, λ odd, and k even, let G be the $(v+1) \times (2v+2)$ matrix whose first $v+1$ columns constitute the identity matrix, whose $(v+2)$-column consists of a 0 in the first row and 1's elsewhere, and whose last v columns are those of M bordered above by a row of 1's. Then the row space of G over F_2 is a $(2v+2, v+1)$ self-dual code.

Examples: For non-trivial examples of Method 1 one can take any one of the three (16,6,2)-designs. For non-trivial examples of Method 2 one can take for the symmetric design any projective plane, the relevant primes being those dividing the order of the plane. Or one can take the (11,5,2) design and produce the celebrated ternary (12,6) Golary code. For a non-trivial example of Method 4 take the symmetric (11,6,3)-design; here one obtains the celebrated (24,12) binary Golay code.

We now restrict our attention to Method 3.

Theorem: Let M be the incidence matrix of a biplane (i.e. a $(v,k,2)$-design) with k odd. Let G be the $v \times 2v$ matrix whose first v columns constitute the identity matrix and whose last v columns are the columns of M. Let G' be the $v \times 2v$ matrix whose first v columns are those of M^t and whose last v columns constitute the identity matrix. Then the mod 2 spans of the row spaces of G and G' are identical and this subspace of F_2^{2v} is a $(2v,v)$ self-dual binary code with minimum weight $k+1$. Moreover, the minimum-weight vectors which are neither the rows of G nor the rows of G' each have $\frac{1}{2}(k+1)$ 1's on the first v coordinates (and hence $\frac{1}{2}(k+1)$ 1's on the last v coordinates also).

Proof: That G and G' have the same row space mod 2 follows from the fact that MM^t is the identity matrix (mod 2) and the standard computation of matrix inverses via row reduction. Call the common subspace generated by G or G' A. That $A \subset A^{\perp}$ is immediate from the fact that any two rows of G have two ones in common, since M is the incidence matrix of a biplane. Since $\dim A = v$, $A = A^{\perp}$ and A is self-dual. Since each row of G has weight $k+1$, the minumum weight of A is at most $k+1$. We next will show that the mod 2 sum of any s distinct rows of G for $2 \leqq s \leqq \frac{1}{2}(k-1)$ is a vector of weight greater than k+1 and since the proof will only involve the fact that M is the incidence matrix of a biplane the same will be true for G', M^t being the incidence matrix of the dual biplane. But any vector has at least half of its ones either in the first v coordinates or the last v and hence we will have proven both that the minimum weight is k+1 and the last assertion of the Theorem concerning the minimum-weight vectors.

Now, for s = 1 the weight is k+1 and for s = 2, the weight is 2k-2. We show that the mod 2 sum of s rows has weight at least s(k+3-2s) by induction on s. For s = 1 or 2 the bound is exact. Assume it is true for s. Any further distinct row has at most 2s 1's in common with a vector which is the sum of s distinct rows since M is the incidence matrix of a biplane. Hence the weight of the sum of s+1 distinct rows is at least s(k+3 -2s) + k+1 - 4s = (s+1)(k+3-2(s+1)), completing the induction. Since s(k+3-2s), as a function of s, is a parabola with its maximum at $\frac{1}{4}(k+3)$, the value for $s \leqq \frac{1}{2}(k+1)$ is at least as great as its value at s = 1, thus completing the proof of the Theorem.

Remark: Whenever $k \equiv -1 \pmod 4$, the generator matrix G has all of its rows of weight congruent to 0 mod 4 and hence we obtain a doubly-even self-dual code. This yields an elementary proof of the fact that there do not exist biplanes with $k \equiv 7 \pmod 8$.

The Examples: There are only finitely many known biplanes. For odd k there are ten known; we next examine the codes obtained from these biplanes.

k = 3. The unique biplane consists of the four 3-subsets of a 4-set. The doubly-even self-dual (8,4) code obtained must of necessity be the extended Hamming code and hence the weight-4 vectors constitute the extension of the projective plane of order 2.

k = 5. Here again the biplane is unique; it is the (11,5,2) quadratic-residue design. We thus obtain a (22,11) code with minimum weight 6. In fact the minimum-weight vectors form a design of type S(3,6,22), i.e. the extension of a projective plane of order 4. We give the very easy proof of this fact and this construction yields, we believe, the most elementary construction of this extension.

First of all no two distinct vectors of weight 6 can have three 1's in common since by orthogonality they would of necessity have at least four 1's in common and then their sum would be a vector of weight less than 6, a contradiction. To finish the proof we directly count that there are at least 77 weight-6 vectors: The rows of G and G' yield 22 such vectors, so we need only show that there are at least (and hence precisely) 55 vectors with three 1's in their first eleven coordinates and three in the last eleven. The sum of any two rows of G is a weight-8 vector. The number of further rows which when added to this vector would yield a weight-6 vector is easily seen to be 3. Hence there are

$\frac{1}{3}[\binom{11}{2} .3] = 55$ such weight-6 vectors.

Observe that this construction not only yields the extension of the plane of order 4 but some interesting facts about its structure, namely that there is a splitting of the 22-set into two 11-sets with all blocks all of the form 1-5, 5-1, or 3-3 and that the automorphism group of the (11,5,2)-design, $PSL_2(F_{11})$, appears as a subgroup of the automorphism group, M_{22}, of the extension.

The success of these elementary constructions certainly whets one's appetite for the consideration of designs of type (56,11,2). To our knowledge there are but two such designs known. We shall postpone discussion of the four biplanes of type (37,9,2) and the two known biplanes of type (74,13,2), discussing (56,11,2) first. Since the circle of ideas being presented allows an elementary construction of one of the known biplanes we sketch that construction first.

Consider the (22,11) code A constructed above whose minimal-weight vectors constitute the S(3,6,22) design. For this design $r = 21$ and the 21 weight-6 vectors through a given point yield, of course, $P_2(F_4)$, the projective plane of order 4. There are then 56 weight-6 vectors with a 0 at some fixed coordinate. Let $\mathbb{B}$ be the set consisting of these fifty-six vectors. Any two distinct such vectors have either zero or two 1's in common. Define a matrix $(M_{B,C}) = M$ with rows and columns indexed by $\mathbb{B}$ via

$$M_{B,C} = 0 \text{ if } |B \cap C| = 2,$$

$$M_{B,C} = 1 \text{ otherwise.}$$

Since $\mathbb{B}$ forms an $S_4(2,6,21)$ and $3\binom{6}{2} = 45$ each row of M

has eleven 1's. We next show that any pair of distinct rows have precisely two 1's in common thus completing the construction: Suppose B and B' are distinct elements of $\mathbb{B}$ viewed as row indices. Case 1, $B \cap B' = \phi$. Then for no C in $\mathbb{B}$ is $B \cap C = \phi = B' \cap C$ for then A would have a vector of weight 18 and hence a vector of weight 4 by complementation, the all-one vector being in A by virtue of the fact that $A = A^{\perp}$. Thus $M_{B,B} = M_{B,B'} = 1 = M_{B',B} = M_{B',B'}$ but for $C \neq B,B'$, $M_{B,C} = 1$ implies $M_{B',C} = 0$. Case 2, $|B \cap B'| = 2$. Here clearly $M_{B,B'} = 0 = M_{B',B}$. Now $M_{B,C} = 1 = M_{B',C}$ if and only if $B \cap C = \phi = B' \cap C$. Of the fifty-four remaining blocks, C, two contain $B \cap B'$ and therefore no remaining points of $B \cup B'$, twenty-four contain precisely one point of $B \cap B'$ and hence one other point of B and one other of B'. There are eighteen blocks meeting $B-B'$ twice and eighteen meeting $B'-B$ twice. Counting flags of the type (p,p',C) where $p \in B - B'$, $p' \in B' - B$ and $\{p,p'\} \subset C$, $C \neq B,B'$ yields $4 \cdot 4 \cdot 4$ on the one hand and $24 + 4x$ on the other where x is the number of C's with $|C \cap (B-B')| = 2 = |C \cap (B'-B)|$. Hence $x = 10$ and there are

$$2 + 24 + 10 + 8 + 8 = 52$$

blocks other than B and B' that meet at least one of them and hence 2 that meet neither. This completes the proof.

Observe that M is a symmetric matrix and one gets free that the constructed (56,11,2) is self-dual. This construction is essentially the one given by Jonsson in [5] and hence the biplane here produced is that biplane. Another elementary construction of this design can be found in [2]. One other (self-dual) (56,11,2)-design was discovered by the Lehigh computer.

$k = 11$. The (56,11,2)-design just constructed has a transitive automorphism group isomorphic to $PSL_3(F_4)$ and is self-dual, [3]. We show that these facts imply that the (112,56) doubly-even, self-dual code A of minimum weight 12 given by the Theorem does NOT yield an extension of a projective plane of order 10.

As we've seen above the incidence matrix M of the (56,11,2)-design can be taken so that $M = M^t$. Thus A possesses an automorphism of order 2 interchanging the first 56 coordinates and the last 56; moreover $PSL_3(F_4)$ acts transitively on the first 56 coordinates and on the last 56. Thus the automorphism group of A acts transitively on the 112 coordinates and has an order divisible by 32 (in fact by a still greater power of 2). But since 112=16·7 the stability group of a point (which would be a subgroup of the collineation group of the plane of order 10) would be of even order, an impossibility by the well-known result of Hughes [4]. Thus, roughly speaking, the known biplane has too many symmetries to yield the sought for S(3,12,112). The Lehigh computer checked that the other known (56,11,2)-design does not yield a projective plane of order 10 either.

Of course there may be other (56,11,2)-designs and should one arise one would be duty bound to test this construction on it. One possible way is to investigate whether or not A has weight-16 vectors. If A has a transitive automorphism group and yields an S(3,12,112) design it cannot have weight-16 vectors because of the result of MacWilliams, Sloane, and Thompson [7]. (The A we have constructed does in fact have weight-16 vectors and a search of several minutes on the Lehigh computer found them.) This is a rather simple first test of whether or not a constructed (112,56) code is combinatorially interesting.

k = 9. There are precisely four biplanes with k=9. One is the difference set defined by the biquadratic residues mod 37. Since the symmetric designs defined by difference sets have transitive automorphism groups and are isomorphic to their duals the (74,37) code produced by the theorem has a transitive automorphism group. Thus, if the weight-10 vectors with a 1 at a given point yield a projective plane of order 8 on the remaining seventy-three coordinates, the same would be true of each of the seventy-four coordinates and one would have an extension of the projective plane of order 8. But the projective plane of order 8 has no extension. Thus this biplane does not involve the projective plane of order 8.

The biplane with k=9 and $PSL_2(8)$ as a group of automorphisms is more interesting. We use the description given by Cameron in [2]. The points of this biplane consist of $P = P_1(F_8)$ and the twenty-eight subgroups of order 3 of $PGL_2(F_8)$. The blocks are indexed by P and the thirty-six 2-subsets of P. The points of the block indexed by P are simply the points of P and the points of a block indexed by a 2-subset of P consist of that 2-subset and the seven subgroups of order 3 with the property that the 2-subset is in one of the three orbits (each of cardinality 3, of course) of that subgroup's action on P. Let M be the incidence matrix of this biplane, the first row being the one indexed by P. In the code generated by $G = (I_{37}|M)$ consider the weight-10 vectors with a 1 at their first coordinate. One has one such vector from G and nine from G'. Sixty-three vectors (which necessarily split 5-5 by the Theorem) are obtained as follows: For each of the sixty-three elements of order 2 of $PGL_2(F_8)$ take the (mod 2) sum of the rows indexed by P and the four 2-subsets constituting the orbits of cardinality 2 of the element. These vectors are all of weight 10 with a 1 at the first coordinate since the row indexed by P has been included. One can show that these above described seventy-three vectors constitute a projective plane of order 8 on coordinates 2 through 74. It then follows that A, the row space of G, is contained in the orthogonal of the row space of the incidence matrix of this plane bordered on the left by a column of 1's and hence all other vectors of weight 10 in A have a 0 at the first coordinate and constitute a certain collection of ovals of the plane of order 8.

The Lehigh computer found a fourth (self-dual)biplane with k=9. The projective plane of order 8 is not involved with this biplane. Note that the one yielding a projective plane of order 8 is not self-dual.

k = 13. The two known biplanes with k=13 are duals of one another and were found by Aschbacher [1]. Since our construction treats a biplane and its dual symmetrically we have just one code A involved here. Although A does contain some vectors of weight 14 that split 7-7 there are not enough such to produce a projective plane of order 12. Electronic computation on the Lehigh computer has verified that no matter which one of the coordinate places is chosen the weight-14 vectors with a 1 at this coordinate do NOT constitute a projective plane of order 12 on the remaining one-hundred-and-fifty-seven coordinates.

Remarks 1. It seems somewhat unreasonable to think that all the biplanes with k=11 and k=13 have been found. Because of the possible connection with putative planes of orders 10 and 12 it might be useful to undertake a systematic search for such designs. 2. The two biplanes yielding projective planes, namely the (11,5,2) and the non-self-dual (37,9,2) can be viewed in the same light as arising from the desarguesian projective planes of orders 4 and 8 via known actions of $PSL_2(4)$ and $PSL_2(8)$ on these planes [6]. This known action is general: $PSL_2(2^r)$ acts on the projective plane of order 2^r with the points in three orbits of size $2^{r-1}(2^r+1)$, 2^r+1, and $2^{r-1}(2^r-1)$. The union of the orbits of size 2^r+1 and $2^{r-1}(2^r-1)$ hold the biplane for $r=2$ and 3. Unfortunately, for $r=4$ one does not get a biplane and presumably only the desarguesian planes of orders 4 and 8 are involved with biplanes.

REFERENCES

1. M. Aschbacher, "On collineation groups of symmetric block designs," J. Combinatorial Theory, 11(1971), 272-281.
2. P. J. Cameron, "Biplanes," Math. Zeitsch. 13(1973), 85-101.
3. M. Hall, Jr., R. Lane, D. Wales, "Designs derived from permutation groups," J. Combinatorial Theory, 8(1970), 12-22.
4. D. R. Hughes, "Generalized incidence matrices over group algebras," Ill. J. Math. 1(1957), 545-551.
5. W. Jonsson, "The (56,11,2) design of Hall, Lane, and Wales," J. Combinatorial Theory, 14(1973), 113-118.
6. H. Lüneburg, "Charakterisierungen der endlichen desarguesschen projektiven Ebenen," Math. Zeitsch 85(1964),419-450.
7. F. J. MacWilliams, N. J. A. Sloane, J. G. Thompson, "On the existence of a projective plane of order 10," J. Combinatorial Theory, 14(1973), 66-78.

OVALS FROM THE POINT OF VIEW OF CODING THEORY

E. F. Assmus, Jr. and Howard E. Sachar

Department of Mathematics, Lehigh University
Bethlehem, Pennsylvania and IBM Corporation

In this short note we use the term <u>oval</u> in its usual sense for a projective plane of odd order but for a projective plane of even order n we mean $n+2$ points no three of which are collinear. We split the discussion into three cases: Case 1 is that of a projective plane of even order congruent to 2 modulo 4. Case 2 is that of a projective plane of even order congruent to 0 modulo 4. Case 3 is that of a projective plane of odd order.

<u>Case 1</u>. If M is the incidence matrix of a projective plane of even order congruent to 2 modulo 4 and A is the row space of M over F_2, the field with 2 elements, then $A \supset A^{\perp}$ and the vectors of weight $n+1$ in A are the lines of the plane and the vectors of weight $n+2$ in A (actually in $A^{\perp}$) are the ovals of the plane. These results are proved by easy geometric arguments; see, for example [1]. In the one such plane known, namely the plane of order 2, the dimension of A is 4. Besides the zero vector and the all-one vector the vectors of A are the seven lines and the seven ovals.

<u>Case 2</u>. If M is the incidence matrix of a projective plane of even order congruent to 0 modulo 4 and A is the row space of M over F_2, let B be the even-weight subcode of A. It is easy to see that B is of codimension 1 in A and that $B = A \cap A^{\perp}$, [2]. Now B is generated by the complements of the lines of A and every vector in B is of

M. Aigner (ed.), Higher Combinatorics, 213-216.

weight congruent to 0 modulo 4. The vectors of weight $n+1$ in A are the lines of the plane and the vectors of weight $n+2$ in $A^{\perp}$ are the ovals. It follows that no oval can be in B. Hence, should one find a plane of even order n congruent to 0 modulo 4 with A of dimension $\frac{1}{2}(n^2+n+2)$ one would have $B = A^{\perp}$ and the plane would have no ovals. No such plane is known.

Before discussing the last case we remark that the seemingly correct object to consider is not the row space over an appropriate finite field but the integral lattice generated by the incidence matrix M, that is the row space of M over Z, the ring of integers. As in the previous cases, the ovals appear in $A^{\perp}$; more precisely we settle Case 3 in the following

<u>Theorem</u>: For a projective plane of odd order n let A be the row space of its incidence matrix over Z and suppose L is an oval. Set $N = n^2+n+1$ and define a vector v_L in Z^N by

$$v_L(P) = \begin{cases} 0 & \text{if } P \in L \\ -1 & \text{if } P \text{ is an interior point of } L \\ 1 & \text{if } P \text{ is an exterior point of } L \end{cases}$$

Now, for each line ℓ of the plane let $v_\ell \in Z^N$ be the characteristic function of ℓ. Then

$$\langle v_\ell, v_L\rangle = n \quad \text{if } \ell \text{ is tangent to } L$$

$$\langle v_\ell, v_L\rangle = 0 \quad \text{if } \ell \text{ is either a secant or an}$$

exterior line of L. Here, $\langle \ldots, \ldots\rangle$ denotes the usual inner product.

Conversely, suppose $v \in Z^N$ has weight n^2 with $\frac{1}{2}n(n+1)$ 1's and $\frac{1}{2}n(n-1)$ -1's with $\langle v, v_\ell\rangle \equiv 0 \pmod{n}$ for all lines ℓ of the plane. Then $v = v_L$ for some oval L.

<u>Proof</u>: The first part of the Theorem has an easy geometric proof. We prove only the converse.
Set $S = \operatorname{Supp} v = \{ P \mid v(P) = \pm 1\}$ and set $L = \{P \mid v(P) = 0\}$. Suppose $P \in L$ and let ℓ be a line through P. Since $\langle v_\ell, v\rangle \equiv 0 \pmod{n}$ either $\ell \cap L = \{P\}$ with $\langle v_\ell, v\rangle = \pm n$ or $|\ell \cap L|$ is even with $\langle v_\ell, v\rangle = 0$. Let $\ell_1, \ldots, \ell_m$ be the

lines through P with $|\ell \cap L|$ even and set $k_i = \frac{1}{2}(n+1 - |\ell_i \cap L|)$. Set $m_+ = |\{\ell \mid \ell \cap L = \{P\}$ and $\langle v_\ell, v\rangle = n\}|$, $m_- = |\{\ell \mid \ell \cap L = \{P\}$ and $\langle v_\ell, v\rangle = -n\}|$. Then we have that

$$nm_+ + \sum_{i=1}^{m} k_i = \frac{1}{2}n(n+1)$$

and

$$nm_- + \sum_{i=1}^{m} k_i = \frac{1}{2}n(n-1).$$

Therefore $n(m_+ - m_-) = n$ or $m_+ = m_- + 1$. Moreover, since $n+1 = m + m_+ + m_-$, $m = n-2m_-$.

Now if $m_- = 0$ for each P in L, $m = n$ and each line of the plane meets L at most twice showing that L is an oval. We also conclude that m_+ is always at least 1. But if $m_- > 0$ for some P in L, say P_0, then for $P \in L$, $P \neq P_0$ each line through P meets the lines contributing to the m_+ and m_- of P_0, except the line through P and P_0, and therefore cannot be tangent to L. Hence for P $m_+ = 0$, a contradiction. This completes the proof and we have described the ovals completely in terms of the row space.

Observe that if n is a prime "mod n" means "over F_p" and we can locate the ovals from the complete weight enumerator of the code since $A \supset A^\perp$. It is only fair to remark that these complete weight enumerators are difficult to determine in the odd order case. We know this enumerator only for the plane of order 3 and present it below, where $a X_0^k X_+^\ell X_-^m$ means there are a vectors with k 0's, ℓ 1's and m -1's:

$$13\,X_0^9X_+^4 + 13\,X_0^9X_-^4$$

$$156\,X_0^7X_+^3X_-^3$$

$$78\,X_0^6X_+X_-^6 + 78\,X_0^6X_+^6X_- + 234\,X_0^6X_+^3X_-^4 + 234\,X_0^6X_+^4X_-^3$$

$$13\,X_0^4X_+^9 + 13\,X_0^4X_-^9 + 234\,X_0^4X_+^6X_-^3 + 234\,X_0^4X_+^3X_-^6$$

$$156\,X_0^3X_+^7X_-^3 + 156\,X_0^3X_+^3X_-^7 + 234\,X_0^3X_+^4X_-^6 + 234\,X_0^3X_+^6X_-^4$$

$$78\,X_0X_+^6X_-^6$$

$$X_+^{13} + X_-^{13} + 13\,X_+^9X_-^4 + 13\,X_+^4X_-^9\ .$$

Since for desarguesian planes of odd order the ovals are classical conics there is no new information here. We only note that the vectors of weight 9 constitute precisely the affine subplanes and the ovals.

REFERENCES

1. E. F. Assmus, Jr., Harold F. Mattson, Jr., and Marcia Guza, "Self-orthogonal Steiner systems and projective planes," Math. Zeitsch. 138)1974), 89-96.
2. Howard E. Sachar, "Error-correcting codes associated with finite projective planes," Ph.D. Thesis, Lehigh University, Bethlehem, Pennsylvania, 1973.

PERMUTATION GROUPS ON UNORDERED SETS

Peter J. Cameron

Merton College, Oxford OX1 4JD, England

Given a permutation group G on a set X, there is a natural action of G on the set X(t) of t-subsets of X. What are the implications for G of assuming some property (such as transitivity or primitivity) for this action, and how is this related to the more familiar concepts such as multiple transitivity and primitivity of G? Such questions are considered in these lectures. Rather than a complete survey, they are a description of some of the many combinatorial methods used in studying these questions, and the combinatorial results and problems that arise; these include Ramsey's theorem, colouring problems, order relations, perfect codes, parallelisms, and Steiner systems. In the first lecture we consider transitivity and (briefly) sharp and generous transitivity, while primitivity is the subject of the second. Other properties which have been studied (such as fractional transitivity and strong primitivity) will not be mentioned.

1. TRANSITIVITY

1.1. Introduction

Recall that a group G is transitive on X if, for all $x,y \in X$, some element $g \in G$ maps x to y. We say G is t-transitive on X if it is transitive on the set of ordered t-tuples of distinct elements of X, and is t-homogeneous (or t-set transitive) if it is transitive on X(t). Clearly both concepts reduce to transitivity when t = 1. The first has received more attention from group theorists because it is inductive, in the sense that G is t-transitive if and only if it is transitive and the subgroup G_x fixing a point x is (t-1)-transitive on the remaining points. We have a trivial implication:

M. Aigner (ed.), Higher Combinatorics, 217-239.

(1.1.1) A t-transitive group is t-homogeneous.

In the next two sections, we shall discuss the converse problem: which groups are t-homogeneous but not t-transitive? The pioneering work on this question for finite groups is the important paper of Livingstone and Wagner [29]; this, with subsequent work of Kantor [23], provides a complete answer. We shall describe their results and recent extensions of them to infinite groups. In §1.2, a relation between orbits of a group on X(s) and X(t) is described; more generally, if the s-subsets of X are coloured in some way, we look at "colour schemes" of t-subsets. §1.3 considers those infinite permutation groups (like the group of order-preserving permutations of $\mathbb{R}$) which are t-homogeneous for all t but not s-transitive for some s. Such a group can be at most 3-transitive; we outline two proofs of this, due to Cameron and Higman. In the final section, we discuss a strengthening of transitivity called "generous transitivity", and the corresponding unordered concept. This leads to an "interchange property" for finite groups, and so back to the colouring problem of §1.2. "Sharp transitivity" is also considered briefly.

1.2. Livingstone and Wagner's Theorem 1

The first results in the paper of Livingstone and Wagner [29] are the following theorem and corollary, under the assumption that X is finite:

(1.2.1) Let G be a permutation group on a set X, with $|X| \geq 2t$. Then G has at least as many orbits on X(t) as on X(t-1).

(1.2.2) With the same hypotheses, if G is t-homogeneous then G is (t-1)-homogeneous.

Their proof involved the character theory of the symmetric group. Soon afterwards, a short number-theoretic proof of (1.2.2) was published by Wielandt [47]. His argument gives the same conclusion under the weaker hypothesis that $|X| \geq t+p^a-1$ for all primepowers p^a dividing t. (In particular, a 6-homogeneous group of degree 8 is 5-homogeneous, that is, a 2-homogeneous group of degree 8 is 3-homogeneous!) Also, despite the title of his paper, Wielandt's argument is valid for infinite groups as well. Another proof for infinite (or sufficiently large finite) X was given by Bercov and Hobby [5], using Ramsey's theorem [37]. Again their result is more general: if G and H are groups such that every H-orbit in X(t) is contained in a G-orbit, then every H-orbit in X(t-1) is contained in a G-orbit. (1.2.2) follows on choosing for H the symmetric group.

Let us return to the original proof. (1.2.1) follows from the assertion that the permutation character of G on X(t-1) is contained in the character on X(t); this holds for G since it holds for the symmetric group. Kantor [24] remarked that the character-theoretic statement follows from the fact that the incidence matrix of X(t-1) against X(t) (with incidence = inclusion) has rank equal to the number of rows. We shall see that, properly formulated, this holds in the infinite case as well, and that it yields a more general combinatorial fact which implies (1.2.1). We give the argument in a very general form.

Let X and Y be sets, and $\rho \subseteq X \times Y$ a relation between them. For $y \in Y$ put $\rho^*(y) = \{x \in X \mid (x, y) \in \rho\}$. If the elements of X are coloured with m colours $c_1, \ldots, c_m$, the colour scheme of an element $y \in Y$ is the m-tuple $(a_1, \ldots, a_m)$, where $a_i = |\rho^*(y) \cap c_i|$. For any set X, $\mathbb{Q}^X$ denotes the $\mathbb{Q}$-vector space of functions from X to $\mathbb{Q}$ (with pointwise operations).

(1.2.3) Suppose X and Y are sets, and $\rho \subseteq X \times Y$ a relation satisfying

(i) $\rho^*(y)$ is finite for all $y \in Y$;

(ii) the linear transformation $\theta : \mathbb{Q}^X \to \mathbb{Q}^Y$ defined by

$$(f\theta)(y) = \sum_{x \in \rho^*(y)} f(x)$$

has kernel 0. If the elements of X are coloured with m colours (all of which are used), then at least m colour schemes of elements of Y occur.

Proof. We may suppose that m is finite. Let c_i be the ith colour and f_i its characteristic function, for $1 \leq i \leq m$; let $\sigma_j = \{a_{1j}, \ldots, a_{mj}\}$ be the jth colour scheme and g_j its characteristic function, for $1 \leq j \leq n$. Then $(f_i\theta)(y) = a_{ij}$ if y has colour scheme σ_j, that is,

$$f_i\theta = \sum_{j=1}^{n} a_{ij}g_j.$$

Since $\{f_1, \ldots, f_m\}$ and $\{g_1 \ldots, g_n\}$ are linearly independent sets, the matrix $A = (a_{ij})$ represents a restriction of θ; so $\mathrm{rank}(A) = m \leq n$.

The hypotheses of (1.2.3) have been verified for a number of relations. We spell out the result in a particularly important case.

(1.2.4) Suppose $0 \leq s \leq t$, $s+t \leq |X|$, and the s-subsets of X are coloured with m colours. Then at least m colour schemes of t-subsets occur.

The fact that the inclusion relation between X(s) and X(t) satisfies (ii) of (1.2.3) under these hypotheses is well known for finite sets X, and its extension to infinite sets is easy (see Cameron [9]). (1.2.1) is an immediate corollary of (1.2.4). Other relations for which (1.2.3) holds are the inclusion relations between s-subsets and blocks in a 2s-design (Ray-Chaudhuri and Wilson [38]), subspaces of finite projective and affine spaces (Kantor [24]), and totally isotropic subspaces of finite polar spaces (Lehrer [27]).

There are two unsolved problems associated with (1.2.4):

Question 1. Describe the colourings of X(s) for which equality holds in (1.2.4).

This question is not precisely put, since it is not clear what sort of answer we want. Here is a sample "satisfactory" answer, in the case $s = 2$, $t = 3$:

(1.2.5) Suppose $|X| \geq 5$, and the edges of the complete graph on X are coloured with m different colours in such a way that only m colour schemes of triangles occur. Then one of the following holds:

(i) there is a colour with valency at most 1 at each vertex;

(ii) $m = 2$, and X is partitioned into two nonempty parts so that edges joining vertices in the same part have one colour, and those joining vertices in different parts have the other;

(iii) $m = 2$, $|X| = 5$, and edges of one colour form a pentagon whose diagonals have the other colour;

(iv) $m = 1$.

Question 2. Suppose the s-subsets of X are coloured with m colours, and let n(t) be the number of colour schemes of t-subsets. Show that n(t) is monotonic increasing for $s \leq t \leq \frac{1}{2}|X|$.

If this question has an affirmative answer, then we may assume $t = s+1$ in Question 1.

1.3. Livingstone and Wagner's Theorem 2

In their book on game theory, von Neumann and Morgenstern ([33], p. 258) asked the question: which finite permutation groups of degree n are t-homogeneous for all integers t with $1 \leq t \leq n$? A game with n players having such a group of symmetries will be fair, in the sense that no subset of the players has any special advantage. Also, such a group preserves no non-trivial collection of subsets of X, and so cannot be studied by combinatorics of the "block design" kind; so it is as well to know all such groups. Their determination was given by Beaumont and Peterson [4]:

(1.3.1) If the permutation group G of finite degree n is t-homogeneous for all $t \leq n$, then either

(i) G is the symmetric or alternating group, or

(ii) G = AGL(1, 5), PGL(2, 5), PGL(2, 8), PΓL(2, 8), n = 5,6,9,9.

The next step was taken by Livingstone and Wagner [29]:

(1.3.2) Let G be a t-homogeneous permutation group on the finite set X, with $|X| \geq 2t$. Then

(i) G is (t-1)-transitive;

(ii) if $t \geq 5$ then G is t-transitive.

This paper contains one of the earliest appearances of an induction argument which has proved of great importance for finite permutation groups. The final stone was laid by Kantor [23]:

(1.3.3) Let G be t-homogeneous but not t-transitive of finite degree $n \geq 2t$. Then one of the following holds:

(i) $t = 2$, $ASL(1, n) \leq G \leq A\Sigma L(1, n)$, $n \equiv 3 \pmod 4$;

(ii) $t = 3$, $PSL(2, n-1) \leq G \leq P\Sigma L(2, n-1)$, $n \equiv 0 \pmod 4$;

(iii) $t = 3$, $G = AGL(1, 8)$, $A\Gamma L(1, 8)$, $A\Gamma L(1, 32)$, $n = 8,8,32$;

(iv) $t = 4$, $G = PGL(2, 8)$, $P\Gamma L(2, 8)$, $P\Gamma L(2, 32)$, $n = 9,9,33$.

The position is very different for infinite groups. The group of all order-preserving permutations of the real numbers is t-homogeneous for all positive integers t (consider piecewise-linear maps) but not even 2-transitive (as no two points can be interchanged). Some of its subgroups share these properties (the piecewise-linear permutations, the permutations of bounded support, etc.). Instead of $\mathbb{R}$ we may use any linearly ordered set in which all open intervals are isomorphic; "large" examples can be constructed as ultraproducts. Further examples include the groups of permutations which preserve or reverse a linear order (2- but not 3-transitive), those which preserve a circular order (2- but not 3-transitive), and those which preserve or reverse a circular order (3- but not 4-transitive). The main theorem of this section asserts that there are no more such groups.

(1.3.4) Suppose the infinite permutation group G is t-homogeneous for all positive integers t, and r- but not (r+1)-transitive for some positive integer r. Then $r \leq 3$, and G preserves a linear or circular order on X or a converse pair of such orders.

We briefly outline two proofs of this theorem. The first, due to Cameron [9], works with finite permutation groups, while G. Higman's proof uses ideas from infinite combinatorics and logic[17].

<u>First proof</u>. Let K_t be the permutation group induced on a t-set by its setwise stabiliser, (so $K_t \leq S_t$, with equality if and only if G is t-transitive). Put $k_t = |K_t|$ and let m_t be the number of orbits of the stabiliser of t points on the remaining points. The number of G-orbits on ordered t-tuples is $t!/k_t$, and the number of orbits on ordered (t+1)-tuples is given by the two expressions

$$(t+1)!/k_{t+1} = (t!/k_t)m_t;$$

we deduce that $k_{t+1} = (t+1)k_t/m_t$. Clearly we have $m_{t+1} \geq m_t - 1$; further analysis shows that in fact $m_{t+1} \geq m_t$ (this involves showing that the stabiliser of t points fixes no further point), with equality only if m_t divides $\frac{1}{2}(t+1)(t+2)$. Thus, if $m_t \geq t+3$, then $m_u \geq u+3$ for all $u \geq t$, whence $k_u \leq (t+2)(t+1)k_t/(u+2)(u+1)$; this is impossible, since the right-hand side approaches zero as $u \to \infty$. Thus we must have $m_t \leq t+2$ for all t. If G is r- but not (r+1)-transitive, then K_{r+1} is a subgroup of index at most r+2 in S_{r+1}. We deduce that either $r \leq 5$, or $K_{r+1} = A_{r+1}$ or S_r. For all except four possibilities, a contradiction results. In the remaining cases, knowledge of K_{r+1} and K_{r+2} gives enough information to apply the characterisations of relations, given below, to obtain the conclusions of the theorem.

Second proof. Let ρ be an orbit of G on ordered (r+1)-tuples of distinct elements. Then ρ is a non-trivial G-invariant (r+1)-ary relation; it is homogeneous, in the sense that its restrictions to any two t-sets are isomorphic, for all t. There are only finitely many (r+1)-ary relations on the set $\{1, \ldots, r+1\}$, say $\rho_1, \ldots, \rho_n$. Take any linear ordering $<$ on X. For each (r+1)-subset S of X, the unique order isomorphism from S to $\{1, \ldots, r+1\}$ (in its natural order) maps ρ to some ρ_i, $1 \leq i \leq n$. By Ramsey's theorem [37], there is an infinite subset Y of X, all of whose (r+1)-subsets yield the same ρ_i. Thus, on Y, ρ is derived from $<$, in the sense that it can be defined in terms of $<$ without quantifiers.

Let a collection E associate with each finite subset T of X a nonempty set E(T) of orders on T, such that E(T) contains the restriction to T of every member of E(T') whenever $T \subseteq T'$. The orders on finite subsets of X from which ρ is derived by the above rule form a collection. ($E(T) \neq \emptyset$, since T is ρ-isomorphic to a subset of Y of the same cardinality.) Zorn's lemma shows that there is a minimal collection E* contained in E, with $|E^*(T)| = 1$ for all T; that is, there is a linear order on all of X from which is derived. (This argument is due to B. H. Neumann, but is similar to Rado [36], Lemma 1. It can also be deduced from the Compactness Theorem for the first-order predicate calculus.)

A theorem of Hodges, Lachlan and Shelah [18] (proved independently by Higman) describes the relations that can occur in this situation. Assuming a transitive automorphism group, any such ρ is derived from a unique relation which is a linear order, the "betweenness" relation induced by a linear order, a circular order, or the "separation" relation induced by a circular order. Without this assumption, some variations on the first two also occur.

The first proof shows more than claimed: there is a function f such that, if G is f(r)-homogeneous and r- but not (r+1)-transitive, then the conclusions of (1.3.4) hold. There is some evidence for an affirmative answer to the following:

Question 3. Show that we can take $f(r) = r+2$.

This is true for $r = 1$: a 3-homogeneous but not 2-transitive infinite group preserves a linear order. This was shown by McDermott [30]. By (1.2.2), such a group is 2-homogeneous; so if λ is a G-orbit on ordered pairs of distinct elements, then exactly one of $(x, y) \in \lambda$, $x = y$, $(y, x) \in \lambda$ holds for any x and y. Take a point x. By the pigeonhole principle, there are distinct y and z with either $(x, y), (x, z) \in \lambda$, or $(y, x), (z, x) \in \lambda$. In either case, and whether or not $(y, z) \in \lambda$, the restriction of λ to $\{x, y, z\}$ is a linear order. Since G is 3-homogeneous, the same holds for any 3-set, and λ is a linear order.

The essential fact here is that the axiomatic definition of a linear order involves only three points. Similar results hold for the other relations in (1.3.4). Thus, betweenness in a linear order and the relation of circular order are both ternary relations whose definitions require four points, while separation in a circular order is a quaternary relation whose definition requires five points. These facts are proved in the "foundations of geometry", since betweenness ane separation are the natural relations on lines in the real Euclidean and projective planes. See, for example, Borsuk and Szmielew [6]. Proofs also appear in Cameron [9]. These facts support the conjecture above; using them, Cameron verified the conjecture for $r = 2$.

A survey of properties of groups acting on ordered sets is given by Holland [19].

It should be noted that there are many examples of groups which are t-transitive for all integers $t \geq 1$. These include the group of all permutations of X whose support has cardinality less than a (where a is infinite and $a \leq |X|$); the group of finitary even permutations; homeomorphism groups of manifolds such as $\mathbb{R}^n$, $\mathbb{R}P^n$, S^n $(n > 1)$; free groups; and certain direct limits of finite symmetric groups. Such groups preserve no non-trivial finitary relations; presumably their study will require topological methods.

1.4. Sharp and generous transitivity

A group G is sharply transitive on a set X if, for any $x, y \in X$, there is a unique $g \in G$ mapping x to y. We say G is sharply t-transitive (resp. sharply t-homogeneous) if it is sharply transitive on the set of ordered t-tuples of distinct elements (resp. on the set of unordered t-subsets). Any group can be represented in a unique way as a sharply transitive permutation group — this is Cayley's theorem. On the other hand, Jordan [21] showed:

(1.4.1) A finite sharply t-transitive permutation group with $t \geq 4$ is the symmetric group S_t or S_{t+1}, the alternating group A_{t+2}, or the Mathieu group M_{11} or M_{12} (with t = 4 or 5 respectively).

This was extended by Tits [42] (and further by Hall [15]):

(1.4.2) There is no infinite sharply t-transitive group with $t \geq 4$.

Zassenhaus [52], [51] found all finite sharply 2- and 3-transitive groups. An important tool is an easy special case of Frobenius' theorem [14], which shows that a sharply 2-transitive finite group is the group $\{x \mapsto ax+b \mid a, b \in F, a \neq 0\}$ of a nearfield F. Some work has been done on the infinite case, usually under assumptions which guarantee the same result: see [32], [42], for example.

From Kantor's theorem (1.3.3) it follows that the only finite sharply t-homogeneous groups are ASL(1, n), $n \equiv 3 \pmod 4$, for t = 2, and AGL(1, 8) and $A\Gamma L(1, 32)$ for t = 3. Some work has been done on the infinite case with extra assumptions, e.g. [12].

Question 4. Show that there is no sharply t-homogeneous infinite group for $t \geq 3$.

A permutation group G is generously transitive on X if, for any two distinct points $x, y \in X$, there is an element $g \in G$ which interchanges x and y. Clearly this is a strengthening of transitivity. More generally, Neumann [34] calls G generously t-transitive if the setwise stabiliser of any (t+1)-set induces on it the full symmetric group S_{t+1}; equivalently, if every G-orbit on unordered (t+1)-tuples corresponds to a single orbit on ordered (t+1)-tuples.

(1.4.3) (i) A group is (t+1)-transitive if and only if it is (t+1)-homogeneous and generously t-transitive.
(ii) A generously t-transitive group is t-transitive.

Proof. (i) This is clear.
(ii) We outline two proofs. The first is by induction. If G is generously t-transitive, then it is transitive and the stabiliser of a point is generously (t-1)-transitive on the remaining points. (Indeed, the converse is also true, so generous t-transitivity is inductive in the sense of §1.1. Clearly any inductive property P(t) implies t-transitivity.) The second proof uses the following lemma of Wagner [45]:

(1.4.4) If G is a group in which the stabiliser of any t-set is transitive on it (and $1 \leq t \leq |X|$), then G is (t-1)-homogeneous.

Proof. Any two (t-1)-subsets whose intersection has size t-2 lie in the same orbit. The result follows by connectedness.

Now if G is generously t-transitive, it is t-homogeneous and generously (t-1)-transitive, whence t-transitive, by (i).

A simple sufficient condition for finite groups is useful:

(1.4.5) If G is t-transitive and the stabiliser of t points has all its orbits on the remaining points of different sizes, then G is generously t-transitive.

Proof. The hypothesis implies that all orbits on ordered (t+1)-tuples have different sizes, whereas orbits corresponding to the same orbit on (t+1)-sets have the same size.

The corresponding unordered concept, generous transitivity on X(t), is a good deal stronger than it first appears. Before considering it, we must make a detour.

A permutation group G, on a set X with $|X| = 2k < \infty$, is said to have the interchange property if, for any partition of X into two k-subsets, there is an element of G interchanging the two parts. Such groups have been studied by Cameron, Neumann and Saxl [11]. There are several interesting examples, including the Mathieu groups M_{12} and M_{24} and the 3- and 4-dimensional affine groups over GF(2).

(1.4.6) Let G be a permutation group on X, with $|X| = 2k$. Then G has the interchange property if and only if, for all even integers t with $0 \leq t \leq k-1$, G has equally many orbits on X(t) and X(t+1).

The proof, like Livingstone and Wagner's proof of (1.2.1), is character-theoretic; but a combinatorial analogue (in the spirit of (1.2.4)) is not known. However, the following observation of Hughes [20] is closely related:

(1.4.7) Let $|X| = 2k$ and $D \subseteq X(k)$; suppose D contains the complement of each of its members. If D is a t-design for some even integer t with $0 \leq t \leq k-1$, then D is also a (t+1)-design.

A first consequence of (1.4.6) is that a group G with the interchange property is transitive. (This is easily proved directly.) A second consequence can be obtained from (1.2.5), since G has equally many orbits on X(2) and X(3): either G is imprimitive, with two blocks of size k or k blocks of size 2, or it is 3-homogeneous. (Imprimitivity is defined in § 2.1.) The imprimitive groups can be completely determined. Clearly an answer to Question 1 would be a great help in answering

Question 5. Determine all groups with the interchange property.

End of diversion.

(1.4.8) Suppose G is generously transitive on X(t), and $|X| \geq 2t$. Then
(i) G is (2t-1)-homogeneous and (2t-2)-transitive on X;
(ii) if $t \neq 3$ or 5, the setwise stabiliser of a (2t-1)-set acts on it as S_{2t-1} or A_{2t-1} (and so G is "almost (2t-1)-transitive").

Proof. The setwise stabiliser of a 2t-set has the interchange property; by (1.4.4), G is (2t-1)-homogeneous. The setwise stabiliser of a(2t-1)-set Y contains an element interchanging any two t-subsets of Y intersecting in a single point. By connectedness, it is (t-1)-homogeneous on Y. (1.2.2) and (1.3.1) now yield (ii), and the rest of (i) can be deduced.

Question 6. Show that generous transitivity on X(t) implies (2t-1)-transitivity (when X is infinite).

2. PRIMITIVITY

2.1. Introduction

Primitivity (like simplicity) is usually defined negatively. A transitive group G on X is imprimitive if there exists a G-invariant equivalence relation on X other than equality and the "all" relation, that is, a G-invariant partition of X into at least two parts each of size at least 2. Such a partition is called a system of imprimitivity, and its members are blocks of imprimitivity (or just blocks). A block can be characterised as a subset of X (other than the empty set, singletons, and X itself) which is equal to or disjoint from all its images under G. A group which is not imprimitive is, of course, primitive. Several further characterisations of primitivity are known (these will not be used here):
(i) the transitive group G on X is primitive if and only if G_x is a maximal subgroup of G;
(ii) the transitive group G on X is primitive if and only if all its orbital graphs are connected (in the weak sense, that is, ignoring directions, though if X is finite it doesn't matter);
(iii) for any group G, the simple objects in the category of G-spaces are just the primitive G-spaces and the trivial two-point space.

We will study three generalisations of primitivity. In all these definitions, G is assumed t-transitive on X. The first two parallel those of §1.1: G is t-primitive if the stabiliser of t-1 points is primitive on the remaining points; G is t-set primitive if it acts primitively on X(t). The final definition takes a negative form, reflecting that of primitivity: G is t-Steiner primitive if it is not an automorphism group of a Steiner system S(t, k, n) on X with $t < k < n$. (A Steiner system S(t, k, n) is a collection of blocks or k-subsets of the n-set X, with the property that any t-subset of X is contained in a unique block.)

All three concepts reduce to primitivity when $t = 1$. (A Steiner system S(1, k, n) is simply a partition of X into k-subsets.) There are two trivial implications:

(2.1.1) If G is either t-primitive or t-set primitive, then G is t-Steiner primitive.

Proof. If G preserves a Steiner system S(t, k, n) on X with $t < k < n$, B is a block of this system, and $x_1, \ldots, x_{t-1} \in B$, then $B - \{x_1, \ldots, x_{t-1}\}$ is a block of imprimitivity for the stabiliser of $x_1, \ldots, x_{t-1}$ acting on $X - \{x_1, \ldots, x_{t-1}\}$, while B(t) is a block of imprimitivity for G acting on X(t).

Question 7. Find all exceptions to the remaining implications among the three concepts, at least for finite groups.

Some work has been done by O'Nan (unpublished), Atkinson [2] and Praeger [35]on 2-Steiner primitive but not 2-primitive groups. Apart from this, the only known result is (2.3.3); the groups occurring in that theorem are all t-primitive but not t-set primitive. Other groups with this property include the symplectic groups Sp(2n, 2) (in both 2-transitive representations) and the group PSL(2, 11) for $t = 2$, and PSL(2, 2^a) (where 2^a-1 is prime) for $t = 3$. The Suzuki groups Sz(q), $q > 2$, are 2-set primitive but not 2-primitive. There are also certain soluble groups which are 2-Steiner primitive but neither 2-primitive nor 2-set primitive.

A 2-transitive group preserves no non-trivial binary relation, and so certainly is primitive. Since both properties are inductive, it follows that

(2.1.2) A (t+1)-transitive group is t-primitive.

The heart of this chapter is the analogous result (2.3.3), the determination of all (t+1)-transitive groups which are not t-set primitive. Such a group is shown to be an automorphism group of a parallelism of X(t). The previous section is purely combinatorial, showing how the binary perfect code theorem of van Lint and Tietävӓinen (see [28]) can be used to give a geometric characterisation of certain parallelisms. (It turns out that all but one of the parallelisms in (2.3.3) arise in this way.) The combinatorial results are also used to characterise certain Steiner systems and biplanes. (It is interesting to compare this connection between biplanes and linear binary codes with the one described by Assmus at this meeting [1].)

The concept of t-Steiner primitivity arises naturally in attempts to generalise "classical" theorems on primitive groups. We consider one example of this, Jordan's theorem on primitive groups having a transitive subgroup of smaller degree. This leads to the problem of determining all Steiner systems whose automorphism groups are transitive on ordered bases. This question is also connected with the preceding section § 2.3. Finally, we describe the relation between primitivity and generous transitivity, and give an application due to Saxl [39] showing how these ideas are used in a specific group-theoretic problem.

2.2. An application of the perfect code theorem

A parallelism of $X(t)$ is an equivalence relation satisfying Euclid's "parallel postulate": for any $x \in X$ and $Q \in X(t)$, there is a unique $Q' \in X(t)$ for which $x \in Q'$ and $Q' \parallel Q$. Alternatively, it is a partition of $X(t)$ into parallel classes, each of which partitions X. For the existence of a parallelism of $X(t)$ with $|X| = n < \infty$, it is clearly necessary that t divides n. It was shown by Baranyai [3] that this condition is also sufficient. His proof involves an application of the Integrity Theorem for flows in networks. See Cameron [10] for a detailed account of parallelisms of $X(t)$, including all the material in this section.

A parallelism of $X(t)$ is said to have the parallelogram property if, whenever the union of two distinct parallel t-subsets is partitioned into two equal parts, these two parts are themselves parallel. The name is taken from the case $t = 2$: if one pair of opposite edges of a quadrilateral are parallel, then so are the other pair.

Examples of parallelisms with the parallelogram property include the usual parallelism of lines in d-dimensional affine space over GF(2), for all $d \geq 1$. (Representing the points by elements of a vector space over GF(2), $\{x_1, x_2\} \parallel \{x_3, x_4\}$ if and only if $x_1 + x_2 = x_3 + x_4$, from which the property is clear.)

(2.2.1) A parallelism of $X(t)$ with the parallelogram property, with $|X| = n < \infty$, is one of the following:

(i) a trivial parallelism with $t = 1$ or $t = n$;

(ii) the relation "equal or disjoint", with $n = 2t$;

(iii) an affine line-parallelism, with $t = 2$, $n = 2^d$;

(iv) a unique example with $t = 4$, $n = 24$.

Proof. We may assume $t > 1$. First we construct an edge-coloured graph. The vertex set is

$$\Omega = \left(\bigcup_{i=0}^{t-1} X(i)\right) \cup Y,$$

where Y is the set of parallel classes. We associate a colour $c(x)$ with each point $x \in X$, and join Q to $Q \cup \{x\}$ (or the parallel class containing $Q \cup \{x\}$ if $|Q| = t-1$) with a $c(x)$-coloured edge whenever $|Q| < t$ and $x \notin Q$.

The graph thus defined has the property that every vertex lies on a unique edge of each colour. (This requires only the definition of a parallelism.) For $x \in X$, let $t(x)$ be the (well-defined and fixed-point-free) permutation on Ω which interchanges the ends of each $c(x)$-coloured edge. It can be checked, using the parallelogram property, that $t(x)$ is a strict automorphism of the coloured graph (mapping any edge to an edge of the same colour). It is now easily shown that the permutations $t(x)$ generate an elementary abelian 2-group T, sharply transitive on Ω. We regard T as the additive group of a vector space over GF(2), and write it in additive notation. Because of the sharp transitivity, we can identify T with Ω in such a way that a set $\{x_1, \ldots, x_i\}$ (or the parallel class containing it if $i = t$) is identified with the element $t(x_1) + \ldots + t(x_i)$ mapping $\emptyset$ to that vertex.

The coloured graph on Ω is then the Cayley diagram of T with respect to the generating set $\{t(x) | x \in X\}$. Any Cayley diagram is associated with a presentation of a group. We consider the presentation of T qua elementary abelian 2-group, that is, what relations hold among the generators $t(x)$ beyond $t(x) + t(x) = 0$ and $t(x) + t(x') = t(x') + t(x)$? To describe these relations, we let V be a vector space over GF(2) with basis $\{v(x) | x \in X\}$, and θ the homomorphism from V to T mapping $v(x)$ to $t(x)$ for all x. The "extra relations" are described by the kernel of θ.

A (binary) error-correcting code is a set of n-tuples of zeros and ones, that is, a subset of a vector space V over GF(2) with a fixed basis. A linear code is a subspace. Decoding a linear code involves evaluating the homomorphism onto the quotient space. We interpret our position with this in mind. We find the element $(v(x_1) + \ldots + v(x_i))\theta$ by starting at $\emptyset$ and following the path with colour sequence $c(x_1), \ldots, c(x_i)$: a colourful and graphic decoding procedure!

A code is perfect e-error-correcting if, for any word v, there is a unique codeword which differs from v in at most e places. Given a code, the extended code is obtained by adding an extra coordinate equal to the sum of all the other coordinated to every codeword; this coordinate is called a parity check. We show that $W = \ker(\theta)$ is an extended perfect (t-1)-error-correcting code. Choose $x \in X$, and let V' and W' be obtained by suppressing the coordinate indexed by x; we must show W' is perfect.

For any $v \in V$, consider the vertex of Ω corresponding to $v\theta$. This is either an i-set $\{x_1, \ldots, x_i\}$ with $i \leq t-1$, or a parallel class containing a unique member $\{x, x_1, \ldots, x_{t-1}\}$ which includes x. In either case $v' - (v(x_1)' + \ldots + v(x_i)') \in W'$ and $i \leq t-1$. To show this element of W' is unique, it is enough to show that a nonzero element of W' has at least 2t-1 entries equal to 1. But an element of W corresponds to a closed circuit with no repeated colours starting at $\emptyset$ in the graph. Such a circuit has length at least 2t; if its length is 2t, the colour sequence is the union of two parallel t-sets.

The perfect code theorem ([41], [28]) asserts, in particular, that a linear perfect e-error-correcting binary code is one of the following:

(i) a trivial code (the zero vector only) with m = e;

(ii) a repetition code (the all-zero and all-one vectors) with m = 2e+1;

(iii) a binary Hamming code (spanned by the characteristic functions of the lines of PG(d-1, 2)), with e = 1, $m = 2^d-1$;

(iv) the binary Golay code (spanned by the characteristic functions of the translates of the quadratic residues mod 23, or of the blocks of the Steiner system S(4, 7, 23)), with e = 3, m = 23.

From this our theorem follows.

We conclude with some applications and related results. A Steiner system S(t, k, n) is said to have the symmetric difference property (SDP) if, whenever two blocks have ½k common points, their symmetric difference is a block. To avoid trivia, we assume k is even and k < 2t. The first interesting case is k = 2t-2, and here we have the result (changing notation slightly):

(2.2.2) The only nontrivial Steiner systems S(t+1, 2t, n) with the symmetric difference property are the system of planes in affine d-space over GF(2) ($d \geq 3$) and the unique S(5, 8, 24).

Proof. Call two t-subsets parallel if either they are equal or their union is a block. That this is an equivalence relation follows from the SDP; that it is a parallelism comes from the definition of a Steiner system; that it has the parallelogram property is trivial.

Question 8. Find all Steiner systems S(t, k, n) (with k even and k < 2t) with the symmetric difference property.

Apart from (2.2.2), little is known. For example, if t = 5 and k = 6 then n = 12 (and the unique S(5, 6, 12) has the SDP). Any further example must have $t \geq 7$ and $k \geq 10$. See [8].

There is an equivalent geometric characterisation of the derived systems S(t, 2t-1, n-1) of systems with SDP, resembling the Veblen-Young axioms [43] for projective geometry. See [10].

A final application concerns biplanes. (We refer to Assmus, Mezzaroba and Salwach [1] for further information on the biplanes discussed here, and Cameron [10] for the connection with parallelisms.) We require two special biplanes: B (order 1) has as blocks all 3-subsets of a 4-set, while a block of B' (order 4) is the symmetric difference of a row and a column in a 4×4 array. In any biplane, three points x, y, z of a block X generate B if and only if the "second blocks" through xy, yz and zx have a common point w. Suppose a biplane has a block X such that any three points of X generate B. Call two 3-subsets of X parallel if the "fourth points" of the biplanes they generate are equal. This is a parallelism of X(3), and the parallelism determines the biplane. Examples of biplanes with this property are B, B', and a unique example of order 7. It is not known whether there are any more. However, under a stronger hypothesis, we can obtain a characterisation:

(2.2.3) Suppose a biplane B* has a block X (with $|X| \geq 4$), any four of whose points generate B'. Then B* = B'.

Proof. Since any three points of a block of B' generate B, we have a parallelism of X(3), which has the parallelogram property (since this holds in B'). Now (2.2.1) applies.

Our discussion so far has assumed that X is finite. In fact everything but the perfect code theorem remains valid for infinite sets X; so linear perfect binary codes, parallelisms with the parallelogram property, Steiner systems S(t+1, 2t, n) with the SDP, and (for t = 3) biplanes satisfying the hypotheses of (2.2.3), are "equivalent" concepts. Arbitrarily large examples are known with t = 2, , constructed as in the finite case; but none is known with $t > 2$.

Question 9. What infinite examples exist?

2.3. Automorphism groups of parallelisms

Let G be an automorphism group of a parallelism of X(t), where $t > 1$ and $|X| = n > 2t$. Then, given t points $x_1, \ldots, x_t$, the t-sets parallel to $\{x_1, \ldots, x_t\}$ form a system of imprimitivity for the stabiliser of $x_1, \ldots, x_t$; so G is not (t+1)-primitive, and *a fortiori* not (t+2)-transitive by (2.1.2). Further, if the parallelism has the parallelogram property, then G is not even (t+1)-Steiner primitive. In this section we discuss two theorems:

(2.3.1) Suppose a parallelism of X(t) (with X finite) admits a (t+1)-transitive automorphism group. Then either it has the parallelogram property, or $t = 2$, $|X| = 6$.

(2.3.2) Suppose the permutation group G on the finite set X is (t+1)-transitive but not t-set primitive. Then G is an automorphism group of a parallelism of X(t), and a block of imprimitivity is a parallel class.

Putting together (2.2.1), (2.3.1) and (2.3.2), we obtain:

(2.3.3) With the hypotheses of (2.3.2), one of the following occurs:

(i) $n = 2t$, $G = S_n$ or A_n;
(ii) $t = 2$, $n = 2^d$, $G \leqslant AGL(d, 2)$;
(iii) $t = 2$, $n = 6$, $G = PGL(2, 5)$;
(iv) $t = 4$, $n = 24$, $G = M_{24}$.

In (ii), it is possible that G is a proper subgroup of AGL(d, 2); there is an example $G = V_{16}A_7$ when $d = 4$, explained by the isomorphy of GL(4, 2) and A_8. Despite work of Wagner [44], D. G. Higman (unpublished), and Kantor [25], it is still an open problem whether any further examples exist.

Historically, (2.3.1) was proved before (2.3.2), and the proof of (2.3.2) requires a detailed knowledge of the groups occurring in the conclusion of (2.3.1). However, a simple statement like (2.3.2) seems to demand a direct proof. Further motivation for seeking such a proof is provided by the following fact and problem.

(2.3.4) A (t+1)-primitive group is t-set primitive.

Question 10. Find the infinite (t+1)-transitive groups which are not t-set primitive; or, at least verify (2.3.2) in the infinite case.

The proof of (2.3.1) is technical. For $t = 2$ it can be done thus. The fixed point set of the stabiliser H of three points x, y, z is a union of 2-sets parallel to $\{x, y\}$, so has even cardinality k. If $k = 4$ then the parallelogram property holds. Techniques developed by Kantor (see [26] chapter 6) show that, if $k > 4$, then $k = n$, that is, G is sharply 3-transitive (see §1.4). From the classification of such groups, we deduce that $n = 6$. The proof in [10] is more elementary. Note that the unique parallelism in the case $t = 2$, $n = 6$ has automorphism group acting as PGL(2, 5) on the points and as S_5 on the colours.

The deduction of (2.3.2) from (2.3.1) is by induction on t. Suppose G is (t+1)-transitive on X, and let S be a block of imprimitivity in X(t). If any two members of S are disjoint, then S is a parallel class. Otherwise, let x be a point contained in two members of S, and let $S' = \{Q - \{x\} \mid x \in Q \in S\}$. S' is a block of imprimitivity for G_x in $(X - \{x\})(t-1)$. By induction, it is a parallel class, whence S is the set of blocks of a Steiner system S(2, t, n) on X.

Furthermore, G is a transitive extension of one of the groups in (2.3.3). Since S_{2t-1} and A_{2t-1} are primitive on t-sets and M_{24} has no transitive extension, we must have t = 3. Let H be the stabiliser of four points of X, and Z its fixed point set. Then $N_G(H)$ is sharply 4-transitive on Z, so $|Z| = 4, 5, 6$ or 11, by (1.4.1). If $x, y, z \in Z$ and $\{x, y, z\} \in S$, then H preserves the Steiner triple system S, and Z carries a subsystem; but there is no Steiner triple system on 4, 5, 6 or 11 points.

An application of the case t = 3, related to (2.2.3):

(2.3.5) A biplane, with a group G of automorphisms fixing a block X and 4-transitive on it, is either B' or the complement of the projective plane over GF(2).

Proof. Suppose $|X| = k > 4$. For $x, y, z \in X$, let $H = G_{xyz}$, and let B_{xy} be the second block through x and y, etc. If $B_{xy} \cap B_{xz} = \{x, w\}$, then w is fixed by H, as is the second block through y and w; this block must be B_{yz}, since its intersection with X is fixed pointwise by H. Thus any three points of X generate B. By the remarks preceding (2.2.3), G preserves a parallelism of X(3), and (2.3.3) applies.

We note in passing that no nontrivial symmetric design with $\lambda > 2$ can admit a group acting 4-transitively on a block. (See [26] section 8.F.)

2.4. Basis-transitive Steiner systems

Of the three concepts introduced in §2.1, t-primitivity has received the most attention from group theorists. This is because the usual way to study a permutation group is via the stabiliser of a point; t-primitivity, like t-transitivity, is inductive. However, a case can be made that t-Steiner primitivity is the most "natural" concept. One piece of evidence is the fact that two classical theorems of Jordan [21] on primitive groups can be extended to t-Steiner primitive groups.

The first theorem asserts that, if G is primitive on X and P is a non-identity Sylow p-subgroup of G_x, then P acts non-trivially on every G_x-orbit different from $\{x\}$. A similar assertion holds for the stabiliser of t points in a t-Steiner primitive group (Sims [40]).

The second concerns what are now known as Jordan groups. If G is a permutation group on a finite set X, a subset Y of X (with $|Y| > 1$ and $|X - Y| \geq 1$) is a Jordan subset if the pointwise stabiliser of X - Y is transitive on Y. A Jordan subset Y is nontrivial if G is not $(|X - Y| + 1)$-transitive. A Jordan group is one admitting a nontrivial Jordan set. Jordan showed that a primitive Jordan group is 2-transitive. (Note that the extension to infinite groups requires topological concepts: Wielandt [48].) More generally,

(2.4.1) A t-Steiner primitive Jordan group is (t+1)-transitive.

Proof. Suppose the Jordan group G is t-transitive but not (t+1)-transitive. It can be shown that any two Jordan sets in $X - \{x_1, \ldots, x_t\}$ have nonempty intersection, so their union is a Jordan set. Then $X - \{x_1, \ldots, x_t\}$ contains a unique maximal Jordan set Y. The translates of $B = X - Y$ are the blocks of a Steiner system on X.

A basis in a Steiner system S(t, k, n) is an ordered (t+1)-tuple whose points are not contained in a block. (Apart from the order requirement, this agrees with the matroid-theoretic definition.) A system admitting a group transitive on bases is called basis-transitive. The proof of (2.4.1) shows that a Jordan group acts on a basis-transitive Steiner system; so the following question generalises the problem of finding all Jordan groups:

Question 11. Find all basis-transitive Steiner systems.

The known non-trivial examples are the systems of points and lines in projective and affine geometries over GF(q) (with $q \neq 2$ for affine geometries), points and planes in affine geometries over GF(2), and the three remarkable systems S(3, 6, 22), S(4, 7, 23), and S(5, 8, 24) discovered by Witt [49], [50].

An answer to this question would have several consequences, beside the classification of Jordan groups. For example, a basis-transitive matroid gives rise, on geometrisation and truncation, to a basis-transitive Steiner system. Also, the following result is related to (2.3.2) and (2.3.4), and is similarly proved:

(2.4.2) Suppose G is (t+1)-transitive but not t-set primitive on X. Then a block of imprimitivity in X(t) is the set of blocks of a basis-transitive Steiner system S(s, t, n) for some $s < t$.

Result (2.3.2) asserts that, for finite sets X, we have $s = 1$. Knowledge of basis-transitive Steiner systems with $s \geq 2$ may help in giving an alternative proof. This result suggests several combinatorial problems:

Question 12. When can the set X(t) be partitioned into Steiner systems S(s, t, n) of some particular type (for example, all isomorphic to projective or affine geometries)?

Question 13. Given a partition of X(t) into Steiner systems S(2, t, n), we have for any $x \in X$ a derived parallelism of

$(X - \{x\})(t-1)$. What is the relation between properties of the Steiner systems occurring in the partition and properties of the derived parallelisms?

We illustrate this question with an example.

(2.4.3) For a finite set X with $|X| = n > 3$, there is no partition of $X(3)$ into Steiner triple systems for which all the derived parallelisms have the parallelogram property.

Proof. (i) By (2.2.1), $n = 2^d+1$ for some integer d.
(ii) Write abc $\sim$ def if these two 3-sets belong to the same Steiner triple system. Given $123 \sim 145 \sim 246$, we have $135 \sim 124 \sim 236$, so $123 \sim 356$. Thus each Steiner triple system is a projective geometry over GF(2) [43], and $n = 2^e-1$ for some integer e.

Combining (i) and (ii) gives the result. (This leaves open the question whether the same holds for infinite sets.) Note that (2.4.3), and Cayley's result that $X(3)$ cannot be partitioned into Steiner triple systems when $|X| = 7$, provide an alternative argument in (2.3.2).

To conclude this section, we look at some of the work that has been done towards answering Question 11.

(2.4.4) A Steiner system $S(t, k, n)$ $(1 < t < k < n)$, admitting a basis-transitive group G in which the stabiliser of t points fixes pointwise the block containing them, is a projective geometry over GF(2) or an affine geometry.

This result is due to Kantor ([22]; see [26] for discussion). Hall [16] had earlier treated the case $t = 2$, $k = 3$, while a result of Nagao [31] shows that the case $t > 3$ cannot occur. I outline Hall's proof; Kantor's follows similar lines. By considering involutions, Hall shows that there is a subsystem S_7 (the projective plane over GF(2)) or S_9 (the affine plane over GF(3)). By basis-transitivity, every basis lies in such a subsystem. In the S_7 case, the Veblen-Young axioms for projective geometry [43] are satisfied. There is a loop L associated with any Steiner triple system, which is commutative and of exponent 3. If every basis lies in an S_9, it is a Moufang loop. Bruck and Slaby ([7] p. 157) showed that such a loop is centrally nilpotent. Since the automorphism group of L is transitive on non-identity elements, every element lies in the centre, and L is an elementary abelian 3-group.

To motivate the next result, note that a group G acts basis-transitively on $S(t, k, n)$ if and only if it is transitive on point-block pairs (x, B) with $x \notin B$, while the stabiliser of such a pair is t-transitive on B.

(2.4.5) Let G act transitively on point-block pairs (x, B) with $x \notin B$ of the Steiner system S(t, k, n) with $2 < t < k < n$, and suppose the stabiliser of such a pair is (t+2)-transitive on B. Then the system is an affine geometry over GF(2) or a Witt system.

Proof. It is sufficient to do the case t = 3. The transitivity assumption implies the following conditions:

(i) if $(x_1, \ldots, x_4)$ is a basis, the number of pairs $\{x_5, x_6\}$ for which $\{x_1, x_2, x_5, x_6\}$ and $\{x_3, x_4, x_5, x_6\}$ are subsets of blocks, is independent of the choice of $(x_1, \ldots, x_4)$;

(ii) either (a) $|B_1 \cap B_2| = |B_1 \cap B_3| = 2$, $B_1 \cap B_2 \cap B_3 = \emptyset$ implies $|B_2 \cap B_3| \neq 1$, for any three blocks B_1, B_2, B_3;

or (b) $|B_1 \cap B_2| = |B_1 \cap B_3| = 2$, $B_1 \cap B_2 \cap B_3 = \emptyset$ implies $|B_2 \cap B_3| \neq 2$, for any three blocks B_1, B_2, B_3.

Condition (ii)(a) implies (i), and implies that the system is a locally projective space in the sense of Doyen and Hubaut [13]; their result shows that it is affine or a Witt system (or S(3, 12, 112) if it exists). In fact, for k = 4, (ii)(a) is just the SDP! The classification can also be done by eigenvalue methods as in [8]. Case (ii)(b) is eliminated group-theoretically; the case k = 5 requires special attention. The combinatorial problem of systems satisfying (i) and (ii)(b) is still unsolved; a brief survey of what is known follows (see [8]).

Condition (ii)(b) for S(3, k, n) implies $n \geq 2 + \frac{1}{2}(k-1)^2(k-2)$, with equality only for the parameters S(3, 5, 26), S(3, 23, 5084) and S(3, 105, 557026). For each of these cases, (ii)(b) implies (i). If (i) and (ii)(b) hold, then k is odd, and if $k > 5$ then n is bounded above by a function of k. (The only possibility for $7 \leq k \leq 21$ is S(3, 7, 2702).) No such bound is known for k = 5; but the existence of S(3, 5, n) satisfying (i) and (ii)(b) implies the existence of a biplane of order n-1 with a null polarity.

2.5 Generous transitivity revisited

We saw in (1.4.1) that generous t-transitivity "lifts" (t+1)-homogeneity to full (t+1)-transitivity. It performs a similar function with respect to primitivity (Neumann [34]):

(2.5.1) A generously t-transitive and t-Steiner primitive group is t-primitive.

Indeed, if G is generously t-transitive and B is any block of imprimitivity for the stabiliser of $x_1, \ldots, x_{t-1}$, then the translates of $\{x_1, \ldots, x_{t-1}\} \cup B$ are the blocks of a Steiner system.

We conclude with an example to show how these concepts can

be used in a specific problem. This application arose in work of Saxl [39] on a theorem of Manning.

(2.5.2) Let G be a 4-transitive group of degree n with the property that the stabiliser of four points has just two orbits on the remaining points, one of them of length 3. Then n = 23, G = M_{23}.

Proof. We may assume, by consulting Sims' list of primitive groups of small degree 40 , that n is sufficiently large. If n > 10 then G is generously 4-transitive by (1.4.5); if n > 13 then G is not 4-primitive, by [40], Theorem 2.1(ii). By (2.5.1), G is an automorphism group of a Steiner system S(4, 7, n), and clearly G is basis-transitive. Since a 4-transitive group of degree 7 is the symmetric or alternating group, (2.4.5) gives the result.

REFERENCES

1. E. F. Assmus Jr., J. A. Mezzaroba and C. J. Salwach, Planes and biplanes. Proc. of this conference.
2. M. D. Atkinson, Doubly transitive but not doubly primitive permutation groups. I, J. London Math. Soc. (2) 7 (1974), 632-634 II, ibid. (2) 10 (1975), 53-60.
3. Z. Baranyai, On the factorisation of the complete uniform hypergraph. Proc. Erdos-Colloq., Keszthely, to appear.
4. R. A. Beaumont and R. P. Peterson, Set-transitive permutation groups. Canad. J. Math. 7 (1955), 35-42.
5. R. D. Bercov and C. R. Hobby, Permutation groups on unordered sets. Math. Z. 115 (1970), 165-168.
6. K. Borsuk and W. Szmielew, "Foundations of Geometry". North-Holland Publ. Co., Amsterdam, 1960.
7. R. H. Bruck, "A Survey of Binary Systems". Springer-Verlag, Berlin-Gottingen-Heidelberg, 1958.
8. P. J. Cameron, Two remarks on Steiner systems. Geometriae Dedicata 4 (1975), 403-418.
9. P. J. Cameron, Transitivity of permutation groups on unordered sets. Math. Z. 148 (1976), 127-139.
10. P. J. Cameron, "Parallelisms of Complete Designs". London Math. Soc. Lecture Notes 23, Cambr. Univ. Pr., Cambridge, 1976.
11. P. J. Cameron, P. M. Neumann and J. Saxl, Finite groups with an interchange property.
12. P. Camion, Sharply two-homogeneous infinite permutation groups. "Permutations" (Actes du Colloq., Paris 1972), 49-55. Gauthier-Villars, Paris, 1974.
13. J. Doyen and X. Hubaut, Finite regular locally projective spaces. Math. Z. 119 (1971), 83-88.
14. G. Frobenius, Über primitive Gruppen des Grades n und der Klasse n-1. S. B. Akad. Berlin (1902), 455-459.

15. M. Hall Jr., On a theorem of Jordan, Pacific J. Math. 4 (1954), 219-226.
16. M. Hall Jr., Group theory and block designs. Proc. Intern. Conf. Theory of Groups (ed. L. G. Kovacs and B. H. Neumann), 113-144. Gordon & Breach, New York, 1967.
17. G. Higman, Homogeneous relations. To appear.
18. W. A. Hodges, A. H. Lachlan and S. Shelah, Possible orderings of an indiscernable sequence. To appear.
19. W. C. Holland, Ordered permutation groups. "Permutations" (Actes du Colloq., Paris 1972), 57-64. Gauthier-Villars, Paris, 1974.
20. D. R. Hughes, Private communication.
21. C. Jordan, "Oeuvres" (ed. J. Dieudonné). Gauthier-Villars, Paris, 1961.
22. W. M. Kantor, On 2-transitive groups in which the stabiliser of two points fixes additional points. J. London Math. Soc. (2) 5 (1972), 114-122.
23. W. M. Kantor, k-homogeneous groups. Math. Z. 124 (1972), 261-265.
24. W. M. Kantor, On incidence matrices of finite projective and affine spaces. Math. Z. 124 (1972), 315-318.
25. W. M. Kantor, on 2-transitive collineation groups of finite projective spaces. Pacific J. Math. 48 (1973), 119-131.
26. W. M. Kantor, 2-transitive designs. "Combinatorics" (ed. M. Hall Jr. and J. H. van Lint), 365-418. Reidel, Dordrecht, 1975.
27. G. I. Lehrer, On incidence structures in finite classical groups. Math. Z. 147 (1976), 287-299.
28. J. H. van Lint, Recent results on perfect codes and related topics. "Combinatorics" (ed. M. Hall Jr. and J. H. van Lint), 163-183. Reidel, Dordrecht, 1975.
29. D. Livingstone and A. Wagner, Transitivity of finite permutation groups on unordered sets. Math. Z. 90 (1965), 393-403.
30. J. P. J. McDermott, lecture at Oxford, 1974.
31. H. Nagao, On multiply transitive groups IV. Osaka J. Math. 2 (1965), 327-341.
32. B. H. Neumann, On the commutativity of addition. J. London Math. Soc. 15 (1940), 203-208.
33. J. von Neumann and O. Morgenstern, "Theory of Games and Economic Behaviour". Wiley, New York, 1964.
34. P. M. Neumann, Generosity and characters of multiply transitive permutation groups. Proc. London Math. Soc. (3) 31 (1975), 457-481.
35. C. E. Praeger, Doubly transitive permutation groups that are not doubly primitive. To appear.
36. R. Rado, Axiomatic treatment of rank in infinite sets. Canad. J. Math. 1 (1949), 337-343.
37. F. P. Ramsey, On a problem in formal logic. Proc. London Math. Soc. (2) 30 (1930), 264-286.
38. D. K. Ray-Chaudhuri and R. M. Wilson, On t-designs, Osaka J. Math. 12 (1975), 737-744.

39. J. Saxl, lecture at Oxford, 1976.
40. C. C. Sims, Computational methods in the study of permutation groups. "Computational Problems in Abstract Algebra" (ed. J. Leech), 169-183. Pergamon Pr., London, 1970.
41. A. Tietäväinen, On the nonexistence of perfect codes over finite fields. SIAM J. Appl. Math. 24 (1973), 88-96.
42. J. Tits, Généralisations des groupes projectifs basée sur leurs propriétés de transitivité. Acad. Roy. Belgique Cl. Sci. Mem. 27 (1952).
43. O. Veblen and J. W. Young, "Projective Geometry". Ginn & Co., Boston, 1916.
44. A. Wagner, On collineation groups of finite projective spaces, I. Math. Z. 76 (1961), 411-426.
45. A. Wagner, Normal subgroups of triply-transitive permutation groups of odd degree. Math. Z. 94 (1966), 219-222.
46. H. Wielandt, "Finite Permutation Groups". Acad. Pr., New York-London, 1964.
47. H. Wielandt, Endliche k-homogene Permutationsgruppen. Math. Z. 101 (1967), 142.
48. H. Wielandt, lecture at Oberwolfach, 1975.
49. E. Witt, Die 5-fach transitiven Gruppen von Mathieu. Abh. Math. Sem. Univ. Hamburg 12 (1938), 256-264.
50. E. Witt, Über Steinersche Systeme. Abh. Math. Sem. Univ. Hamburg 12 (1938), 265-275.
51. H. Zassenhaus, Kennzeichnung endlicher linearer Gruppen als Permutationsgruppen. Abh. Math. Sem. Univ. Hamburg 11 (1935), 17-40.
52. H. Zassenhaus, Über endliche Fastkorper. Abh. Math. Sem. Univ. Hamburg 11 (1935), 187-220.

CODES AND DESIGNS

J.H. van Lint

Department of Mathematics, Technological University,
Eindhoven, Netherlands

1. INTRODUCTION

We denote by $V(n,q)$ the set of all n-tuples from a q-symbol alphabet $\mathbb{F}$ (i.e. $V(n,q) = \mathbb{F}^n$). If q is a prime power we take $\mathbb{F} = \mathbb{F}_q$ and interpret $V(n,q)$ as n-dimensional vector space over $\mathbb{F}_q$. In any case we distinguish an element of $\mathbb{F}$ and denote it by 0. The elements of $V(n,q)$ are called *words* (or vectors) and are denoted by underlined symbols. The word $(0,0,\ldots,0)$ is denoted by $\underline{0}$. The (Hamming) *distance* $d(\underline{x},\underline{y})$ of two words $\underline{x}$ and $\underline{y}$ is defined by

$$d(\underline{x},\underline{y}) := \# \{i \mid 1 \le i \le n, x_i \ne y_i\} .$$

The *weight* $w(\underline{x})$ of the word $\underline{x}$ is $d(\underline{x},\underline{0})$. A subset C of $V(n,q)$ is called a *code*. We say that C is e-*error-correcting* if $d(\underline{x},\underline{y}) \ge 2e + 1$ for all pairs $\underline{x} \in C$, $\underline{y} \in C$, with $\underline{x} \ne \underline{y}$. The minimum distance of C, defined by $d := \min\{d(\underline{x},\underline{y}) \mid \underline{x} \in C, \underline{y} \in C, \underline{x} \ne \underline{y}\}$ determines the error correcting capability $e := \lfloor \frac{d-1}{2} \rfloor$ of the code. The set $S_e(\underline{x}) := \{\underline{y} \in V(n,q) \mid d(\underline{x},\underline{y}) \le e\}$ is called the sphere of radius e around $\underline{x}$. If the set of spheres $S_e(\underline{c})$, where $\underline{c}$ runs through C, forms a partition of $V(n,q)$ then C is called a *perfect code*.

M. Aigner (ed.), Higher Combinatorics, 241-256.

The following numbers play an important role in our investigations:

$$B(\underline{x},k) := \# \{\underline{c} \in C \mid d(\underline{x},\underline{c}) = k\} \tag{1.1}$$

for $\underline{x} \in V(n,q)$, $0 \le k \le n$, i.e. $B(\underline{x},k)$ is the number of code words at distance k from $\underline{x}$. The distance of $\underline{x}$ and C is defined by

$$\rho(\underline{x}) := \min\{k \mid 0 \le k \le n,\ B(\underline{x},k) \ne 0\} . \tag{1.2}$$

In recent years many designs (i.e. t-designs; cf. (2.3)) have been found by considering the words of fixed weight in some special (often linear) code. Well known examples of such codes are the perfect codes (Hamming codes, Golay codes). The codes which have the property display a high degree of regularity. This has led to a number of definitions of classes of codes which could possibly yield unknown designs. One of these is the class of *uniformly packed* codes. These codes were introduced in 1971 by N.V. Semakov, V.A. Zinoviev and G.V. Zaitsev [9]. A few examples and elementary results were given by this author at the previous Advanced Study Institute on Combinatorics in 1974 (cf. [7]). The definition was then generalized and most of the theory necessary to understand these codes was given by J.M. Goethals and H.C.A. van Tilborg in 1975 [6]. In his thesis which appeared a few months ago, H.C.A. van Tilborg [12] gave an extensive treatment of the theory, many examples, but most important of all a proof that if q is a prime power and $e \ge 4$ then such codes do not exist! The main purpose of this paper is to give the reader some idea of these most recent results. For details the reader is referred to [12].

2. REGULAR CODES AND DESIGNS

(2.1) DEFINITION. A code is called t-*regular* if for all $\underline{x} \in V(n,q)$ with $\rho(\underline{x}) \le t$ and for all k $(0 \le k \le n)$ the number $B(\underline{x},k)$ depends only on $\rho(\underline{x})$ and k. If in this definition t is maximal then we call the code *completely regular*.

We shall say that a vector $\underline{x} \in V(n,q)$ is *covered* by a vector $\underline{y} \in V(n,q)$ if for every nonzero coordinate x_i we have $x_i = y_i$.

(2.2) DEFINITION. A q-ary $t-(n,k,\lambda)$ *design* is a collection S of vectors of weight k in $V(n,q)$ with the property that every vector $\underline{x} \in V(n,q)$ of weight t is covered by exactly λ vectors $\underline{y} \in S$.

(2.3) LEMMA. If S is a q-ary $t-(n,k,\lambda)$ design then

i) S is also a q-ary $i-(n,k,\lambda_i)$ design for $0 \le i \le t-1$, where

$$\lambda_i := \frac{\lambda\binom{n-i}{t-i}(q-1)^{t-i}}{\binom{k-i}{t-i}} .$$

ii) If $w(\underline{x}) = i$ and we consider j zero coordinates of $\underline{x}$ where $i+j \le t$ then the number of vectors $\underline{y} \in S$ which cover $\underline{x}$ and which also have zero coordinates at the j prescribed positions depends only on i and j.

PROOF.

i) Let $\underline{x} \in V(n,q)$, $w(\underline{x}) = i$. Then $\underline{x}$ is covered by $\binom{n-i}{t-i}(q-1)^{t-i}$ vectors $\underline{x}'$ of weight t. Each of these vectors $\underline{x}'$ is covered by exactly λ vectors $\underline{y}$ in S. For each vector $\underline{y} \in S$ which covers $\underline{x}$ there are $\binom{k-i}{t-i}$ vectors $\underline{x}'$ of weight t which cover $\underline{x}$ and which are covered by $\underline{y}$.

ii) This follows from i) by a straightforward inclusion-exclusion argument. □

The connection between design theory and coding theory is now established by the following theorem.

(2.5) THEOREM. Let C be a t-regular code with $\underline{0} \in C$. Let the minimum distance d of C satisfy $d \geq 2t$. Then for each k $(0 \leq k \leq n)$ the collection of code words of weight k is a q-ary t-design.

PROOF. The proof is by induction. First consider $k = d$. Let $\underline{x} \in V(n,q)$, $w(\underline{x}) = t$. Since $d \geq 2t$ we have $\rho(\underline{x}) = t$ by the triangle inequality. Now the regularity of C implies that $B(\underline{x}, d-t)$ is independent of $\underline{x}$. Again using the triangle inequality we see that $\underline{c} \in C$ and $d(\underline{x},\underline{c}) = d-t$ iff $w(\underline{c}) = d$ and $\underline{c}$ covers $\underline{x}$. Therefore the words of weight d form a q-ary $t-(n,d,\lambda(d))$ design for some $\lambda(d)$.

Now assume that the theorem has been proved for all $k < w$. Then by lemma 2.3ii) there are numbers $a(k,\ell)$, $(d \leq k < w,\ 0 \leq \ell \leq n)$ such that for any $\underline{x} \in V(n,q)$ with $w(\underline{x}) = t$ exactly $a(k,\ell)$ code words of weight k have distance ℓ to $\underline{x}$. It follows that the number of code words of weight $< w$ with distance $w-t$ to $\underline{x}$ does not depend on the choice of $\underline{x}$. Since we also know that $B(\underline{x}, w-t)$ is independent of the choice of $\underline{x}$ the result again follows from the triangle inequality for Hamming distance. □

3. RECENT RESULTS ON PERFECT CODES

Since perfect codes are completely regular they yield interesting designs and in fact the known non-trivial perfect codes are connected to even nicer designs than theorem 2.5 promises. It has been known for some years that if q is a prime power then there are no unknown perfect codes. The situation for other q is still far from being completely understood. We shall briefly report on the state of affairs at present. We consider perfect codes in $V(n,q)$ with $d = 2e + 1$.

i) For $e = 1$ and $e = 2$ there are some isolated results. The situation is not understood at all.

ii) The case $e \geq 2$, $q = 2^a 3^b$ was settled in 1975 by L.A. Bassaligo, V.A. Zinoviev, V.K. Leontiev and N.I. Feldman [2].

iii) $e \geq 3$, $q = p_1^r p_2^s$ was settled recently by A. Tietäväinen [11].

iv) $3 \leq e \leq 5$, q arbitrary was settled by H.F. Reuvers [8]. He also has some results concerning $e = 2$ and $q = p_1^r p_2^s$ but the problem seems very hard.

v) E. Bannai [1] has just shown that for each $e \geq 3$ there are at most finitely many unknown perfect codes.

In all the cases above no unknown perfect code turned up and hence the results are negative for those readers interested in t-designs.

4. UNIFORMLY PACKED CODES

We now consider a class of codes which in some sense are one step away from being perfect. For the code C in V(n,q) we shall now require that it is e-error-correcting, that every $\underline{x}$ has $\rho(x) \leq e + 1$, and that it is completely regular. We first need a lemma.

(4.1) LEMMA. Let C be an e-error-correcting code in V(n,q). Then for any $\underline{x} \in V(n,q)$ we have

$$\text{i)}\quad \rho(\underline{x}) = e \quad\Rightarrow B(\underline{x}, e+1) \leq \frac{(n-e)(q-1)}{e+1},$$

$$\text{ii)}\quad \rho(\underline{x}) = e+1 \Rightarrow B(\underline{x}, e+1) \leq \frac{n(q-1)}{e+1}.$$

PROOF.

i) Assume w.l.o.g. that $\underline{x} = 0$ and that the code word $\underline{c}$ has $w(\underline{c}) = e$. Then a code word with weight $e+1$ must have 0's in the nonzero positions of $\underline{c}$. Furthermore two such code words cannot have the same nonzero coordinate in the same position. The result follows by double counting.

ii) The proof is analogous. □

(4.2) DEFINITION. An e-error-correcting code C in V(n,q) is called *uniformly packed* with parameters λ and μ, iff for all $\underline{x} \in V(n,q)$

$$\rho(\underline{x}) = e \quad \Rightarrow B(\underline{x}, e+1) = \lambda ,$$

$$\rho(\underline{x}) = e+1 \Rightarrow B(\underline{x}, e+1) = \mu ,$$

where $\lambda < \frac{(n-e)(q-1)}{e+1}$, $\mu \geq 1$.

Clearly the definition implies that $\rho(\underline{x}) \leq e+1$ for all $\underline{x}$. The conditions of (4.2) uniquely determine the number of code words.

(4.3) THEOREM. Let C be an e-error-correcting uniformly packed code in V(n,q) with parameters λ and μ. Then

$$|C|\{\sum_{i=0}^{e-1} \binom{n}{i}(q-1)^i + (1-\frac{\lambda}{\mu})\binom{n}{e}(q-1)^e + \frac{1}{\mu}\binom{n}{e+1}(q-1)^{e+1}\} = q^n. \qquad (4.4)$$

PROOF. The result follows immediately form (4.2) if we count the number of pairs $(\underline{x},\underline{c})$ with $\underline{c} \in C$ and $e \leq d(\underline{x},\underline{c}) \leq e+1$ in two ways. □

Note that if we take $\lambda = \frac{(n-e)(q-1)}{e+1}$ then (4.4) implies that C is perfect. This is why we have excluded this possibility in (4.2). Except for the extended Golay code all known uniformly packed codes have e = 1 or e = 2 (cf. the tables in [6] and [12]). The reader interested in examples is referred to chapter 4 of [12]. Many of the more interesting examples are based on the following theorem which can be proved using the group algebra approach to coding. This would take too along to explain here so we just state the theorem.

(4.5) THEOREM. Let C be an e-error-correcting linear code. Then C is uniformly packed iff in the dual code $C^{\perp}$ exactly $e+1$ nonzero weights occur.

In [4] Delsarte considered n-subsets S of the set of points of a $(k-1)$-dimensional projective space $PG(k-1,q)$, having the property that the intersection of S with any hyperplane consists of $n-w_1$ or $n-w_2$ points. Many such sets S are known. Using such a set S one immediately finds a linear code of length n and dimension k over $\mathbb{F}_q$ having the two weights w_1 and w_2 by considering the characteristic functions of hyperplanes in the space and restricting to S. Delsarte called these codes *two weight projective codes*. By theorem 4.5 the dual of such a code is uniformly packed with $e = 1$. So in this case the codes are constructed using known combinatorial designs.

5. GENERALIZATION OF LLOYD'S THEOREM TO UNIFORMLY PACKED CODES

In the case where q is a power of a prime the group-algebra methods of [6] and [12] led to a generalization of Lloyd's theorem for perfect codes. Recently D.M. Cvetković and J.H. van Lint [3] gave a simple proof of Lloyd's theorem. We shall now show that the same method can be used to prove the generalization to uniformly packed codes for all q. In order to state the theorem we first need a few definitions.

(5.1) DEFINITION. The *Krawtchouk polynomial* K_k is defined by

$$K_k(n,u) := \sum_{j=0}^{k} (-1)^j (q-1)^{k-j} \binom{u}{j}\binom{n-u}{k-j} .$$

The polynomial ψ_e defined by

$$\psi_e(n,x) := K_e(n-1,x-1)$$

is called Lloyd's polynomial of degree e.

Several recurrence relations for Krawtchouk polynomials are known (cf. [10], [5] (4.11)). From these we find

$$(e+1)\psi_{e+1}(n,x) = \{e+(q-1)(n-e)-qx+1\}\psi_e(n,x) - $$
$$+ (q-1)(n-e)\psi_{e-1}(n,x) . \qquad (5.2)$$

The theorem which we wish to prove now follows.

(5.3) THEOREM. Let C be an e-error-correcting uniformly packed code with parameters λ and μ. Then the polynomial Q(x) defined by

$$\mu Q(x) := \psi_{e+1}(n,x) + (\mu-\lambda-1)\psi_e(n,x) +$$
$$+ \lambda\psi_{e-1}(n,x) \qquad (5.3)$$

has $e+1$ distinct integral zeros in the interval $[1,n]$.

The fact that these zeros are distinct follows from well known properties of Krawtchouk polynomials. The interesting fact and the basis of the known non-existence theorems is that the zeros are integers. The proof of theorem 5.3 is based on the following more or less trivial lemma.

(5.5) LEMMA. Let A be a matrix of size m by m which has the form

$$A = \begin{pmatrix} A_{00} & A_{01} & \cdots & A_{0k} \\ A_{10} & A_{11} & \cdots & A_{1k} \\ & & & \\ A_{k0} & A_{k1} & \cdots & A_{kk} \end{pmatrix}$$

where A_{ij} has size m_i by m_j $(i = 0,1,\ldots,k;$ $j = 0,1,\ldots,k)$.

Suppose that for each i and j the matrix A_{ij} has constant row sums with sum b_{ij}. Let the matrix B have entries b_{ij}. Then each eigenvalue of B is also an eigenvalue of A.

PROOF. Let $B\underline{x} = \lambda\underline{x}$, where $\underline{x} = (x_0, x_1, \ldots, x_k)^T$. Define $\underline{y}$ by

$$\underline{y}^T := (x_0, x_0, \ldots, x_0, x_1, x_1, \ldots, x_1, \ldots, x_k, x_k, \ldots, x_k)$$

where each x_i is repeated m_i times. By definition of B it is obvious that $A\underline{y} = \lambda\underline{y}$. □

(5.6) DEFINITION. The square matrix A_n of size q^n is defined as follows. Rows and columns are indexed by the elements of V(n,q) and $A(\underline{x},\underline{y}) = 1$ if $d(\underline{x},\underline{y}) = 1$, $A(\underline{x},\underline{y}) = 0$ otherwise.

From the definition of A_n we see that

$$A_{n+1} = I_q \times (A_n - I_{q^n}) + J_q \times I_{q^n}, \qquad (5.7)$$

where I_m denotes the identity matrix of size m, J_m the all one matrix of size m and × indicates the Kronecker product.

(5.8) LEMMA. The matrix A_n has the eigenvalues

$$-n + jq \quad (j = 0,1,\ldots,n)$$

with multiplicities $\binom{n}{j}(q-1)^j$.

PROOF. The proof is by induction. For n = 1 we have $A_1 = J_q - I_q$ and then the assertion is well known. Suppose $A_n\underline{x} = \lambda\underline{x}$. Let

$$\underline{x}' := (c_1\underline{x}^T, c_2\underline{x}^T, \ldots, c_q\underline{x}^T)^T,$$

where $c_1 + c_2 + \ldots + c_q = 0$. Then by (5.7) $A_{n+1}\underline{x}' = (\lambda - 1)\underline{x}'$. On the other hand, if we take $c_1 = c_2 = \ldots = c_q = 1$ then by (5.7) we find $A_{n+1}\underline{x}' = (\lambda + q - 1)\underline{x}'$. The result now follows from well known addition properties of binomial coefficients. □

Now we assume that C is a uniformly packed code in V(n,q) with parameters λ and μ. We reorder the rows and columns of A_n in such a way that those indexed by an element of C come first, then the rows and columns indexed by elements of V(n,q) with distance 1 to the code, etc. Then A_n has the form of the matrix A in lemma 5.5 with $k = e + 1$ and where A_{ij} has rows (resp. columns) indexed by the elements of V(n,q) with distance i (resp. j) to C. We find (writing $q - 1 =: s$)

$$B = \begin{pmatrix} 0 & ns & 0 & \cdots & & & 0 \\ 1 & q-2 & (n-1)s & 0 & & & \vdots \\ 0 & 2 & 2(q-2) & (n-2)s & & & \\ \vdots & & 0 & \ddots & \ddots & & \\ & & & e-1 & (e-1)(q-2) & (n-e+1)s & 0 \\ & & & & e & e(q-2)+\lambda(e+1) & (n-e)s-\lambda(e+1) \\ 0 & \cdots & & & 0 & \mu(e+1) & ns-\mu(e+1) \end{pmatrix} \qquad (5.9)$$

We now have to determine the eigenvalues of the tridiagonal matrix B. This is done with the same method which was used in [3].

(5.10) DEFINITION. The matrix $Q_e = Q_e(a,b,s)$ is the tridiagonal matrix given by

$$Q_e(a,b,s) := \begin{pmatrix} a & 0 & 0 & 0 & \cdots & 0 \\ 1 & a+(s-1) & b-s & 0 & \cdots & 0 \\ 0 & 2 & a+2(s-1) & b-2s & 0 \cdots & 0 \\ 0 & 0 & \ddots & \ddots & \ddots & \vdots \\ \vdots & & & \ddots & \ddots & b-(e-1)s \\ 0 & \cdots & & 0 & e & a+e(s-1) \end{pmatrix}.$$

Furthermore we define

$$P_e = P_e(a,b,s) := \left(\begin{array}{c|c} Q_{e-1}(a,b,s) & \begin{array}{c}1\\1\\ \vdots\end{array} \\ \hline 0\ 0\ \dots\ 0\ e & 1\end{array}\right) .$$

The determinants of these matrices are denoted by $\bar{Q}_e$, resp. $\bar{P}_e$.

Developing by the last row we find from (5.10)

$$\bar{Q}_e = (a + e(s-1))\bar{Q}_{e-1} - e(b-(e-1)s)\bar{Q}_{e-2} . \qquad (5.11)$$

By adding all columns to the last one we find, developing by the last row

$$\bar{Q}_e = (a+es)\bar{Q}_{e-1} - e(a+b)\bar{P}_{e-1} . \qquad (5.12)$$

Developing P_e by the last row yields

$$\bar{P}_e = \bar{Q}_{e-1} - e\bar{P}_{e-1} . \qquad (5.13)$$

Now apply (5.13) with $e+1$ instead of e, combine with (5.13) and eliminate the $\bar{Q}$-terms using (5.12). This yields

$$\bar{P}_{e+1} = (a+es-e-1)\bar{P}_e - e(b-es)\bar{P}_{e-1} . \qquad (5.14)$$

(5.15) LEMMA. Let $s := q-1$. Then we have

$$\bar{P}_e(qy-ns,ns,s) = (-1)^e e!\psi_e(n,y) .$$

PROOF. For $e = 1$ and $e = 2$ the assertions can be checked directly from the definitions. By substitution of the appropriate values of a and b in (5.14) and using (5.2) we see that the polynomials on both sides in the lemma satisfy the same recurrence relation. □

Now consider the matrix B of (5.9). We shall calculate

$\det((qy - ns)I_{e+2} + B)$. First we add all columns to the last column and take out a factor qy. Now λ and μ occur only in the next to last column. If we replace λ by 0 and μ by 1 then the determinant is $\bar{P}_{e+1}(qy - ns, ns, s)$. Therefore expansion by the next to last column yields

$$qy\{\bar{P}_{e+1} + \lambda(e+1)\bar{Q}_{e-1} - (\mu - 1)(e+1)\bar{P}_e\}$$

and by (5.12) and lemma 5.15 this equals

$$(-1)^{e+1}(e+1)!\,qy\{\psi_{e+1}(n,y) + (\mu - \lambda - 1)\psi_e(n,y) + \lambda\psi_{e-1}(n,y)\} .$$

This expression for $\det((qy - nx)I_{e+2} + B)$ combined with lemmas 5.5 and 5.8 proves theorem 5.3.

6. SKETCH OF NONEXISTENCE PROOFS

Theorem 4.3 and theorem 5.3 are two necessary conditions for the existence of a uniformly packed code C in $V(n,q)$ with parameters λ and μ. We shall now sketch how H.C.A. van Tilborg [12] used these conditions to prove the following theorem.

(5.1) THEOREM. There is no e-error-correcting uniformly packed code with $e \geq 4$ and $q = p^a$ a prime power.

The proof uses the following information about the zeros $x_1, x_2, \ldots, x_{e+1}$ of the polynomial $Q(x)$ defined in (5.4).

$$\sum_{i=1}^{e+1} x_i = \frac{(e+1)}{q}\{(q-1)n + \mu - \lambda - \frac{(q-2)e}{2}\}, \tag{6.2}$$

$$\prod_{i=1}^{e+1} x_i = \frac{\mu(e+1)!}{q^{e+1}}\{\sum_{i=0}^{e-1}\binom{n}{i}(q-1)^i + (1 - \frac{\lambda}{\mu})\binom{n}{e}(q-1)^e + \frac{1}{\mu}\binom{n}{e+1}(q-1)^{e+1}\} , \tag{6.3}$$

$$\prod_{i=1}^{e+1} (x_i - 1) = \frac{(q-1)^{e-1}}{q^{e+1}}(n-1)(n-2)\ldots(n-e+1)p_2(n) \ , \quad (6.4)$$

where

$$p_2(n) = (n-e)(n-e-1)(q-1)^2 + $$
$$+ (\mu-\lambda-1)(n-e)(e+1)(q-1) + \lambda e(e+1) \ .$$

These expressions are obtained easily by calculating a few of the coefficients of Q(x) and Q(1). In a similar way (calculating one more coefficient) an expression for $\sum_{i<j} (x_i - x_j)^2$ can be obtained which then gives an upper bound for the difference between the smallest and largest zero. (This bound has shortened the nonexistence proofs for perfect codes considerably).

Since the Krawtchouk polynomials are orthogonal polynomials the familiar theorems on interlacing of zeros can be used. Let $a_1 < a_2 < \ldots < a_e$ be the zeros of $\psi_e(n,x)$ and let $b_1 < b_2 < \ldots < b_{e+1}$ be the zeros of $\psi_{e+1}(n,x)$. One can then show that

$$0 < x_1 < a_1 < x_2 < a_2 < \ldots < a_e < x_{e+1} \ , \quad (6.5)$$

$$x_{e+1} \le n \text{ iff } \mu - \lambda - 1 \le \frac{n-e-1}{e+1} + \frac{\lambda e}{n-e} \ . \quad (6.6)$$

Furthermore, at least e zeros of Q(x) are in the interval (b_1, b_{e+1}) and the method described above gives the following upper bound for $b_{e+1} - b_e$.

$$f := f(n,q,e) := q(b_{e+1} - b_1) \le$$

$$\le \sqrt{\frac{e(e+1)}{6}\{12n(q-1) + (e+2)q^2 - 12(q-1)(e+1)\}} \quad (6.7)$$

The essential idea of the proof is to derive a lower bound for n and an upper bound for n such that these bounds yield a contradiction for all but finitely many parameter sets n,q,e.

(6.8) LEMMA. If an e-error-correcting, uniformly packed code exists in V(n,q), then

$$n \geq \begin{cases} q^{\frac{e+1}{6}} & \text{for } e \geq 5,\ q \neq 2\ , \\ q.2^{-1/3} & \text{for } e = 4,\ q \neq 2\ , \\ 2^{\frac{e+7}{7}} & \text{for } q = 2\ . \end{cases} \qquad (6.9)$$

PROOF. The proof is based on (6.4). In the proof we shall use the following notation

$$k = T(k).p^{ex(k)}, \quad \text{where } (T(k),p) = 1\ . \qquad (6.10)$$

Let $\ell := \max\{ex(n-k) \mid k = 1,2,\ldots,e-1\}$. We have

$$c := ex((n-1)(n-2)\ldots(n-e+1)) \leq \ell + \frac{e-3}{p-1}\ ,$$

so

$$n \geq p^{\ell} \geq p^{c - \frac{e-3}{p-1}}\ .$$

Let $b := ex(p_2(n))$. From (6.6) it follows that $p_2(n) < q^2n^2$, i.e. $n \geq p^{\frac{b-2a}{2}}$. These two lower bounds for n, combined with the fact that $b + c \geq a(e+1)$, which follows from (6.4), and a few simple calculations establish the first inequality of the lemma. The others are proved in a similar way. □

If $y_1,y_2,\ldots,y_k$ are distinct integers in the interval [1,A] and we assume that they are divisible by the highest possible powers of p a simple counting argument shows that

$$p^{ex(y_1y_2\ldots y_k)} \leq p^{\frac{k-1}{p-1}}\left(\frac{A}{k}\right)^k\ .$$

A similar result holds if $y_1,y_2,\ldots,y_k$ are in an interval $(t,t+A]$ if we exclude one of the integers y_i which could possibly be divisible by a very high power of p.

We now come to the crucial part of the proof, i.e. the upper bound on n. Renumber the zeros of Q(x) in such a way that $x_1,x_2,\ldots,x_e$ are in (b_1,b_{e+1}) and $ex(x_1) \le ex(x_2) \le \ldots \le ex(x_e)$. Then by (6.11) and (6.7) we have

$$p^{ex(x_1x_2\ldots x_{e-1})} \le p^{\frac{e-2}{p-1}}\left(\frac{f}{q(e-1)}\right)^{e-1} .$$

Since b_{e+1} is larger than the largest zero of $\psi_2(n,x)$, which exceeds $\frac{(q-1)n}{q}$, we have

$$x_i \ge \frac{(q-1)n}{q} - (b_{e+1} - b_1) = \frac{(q-1)n-f}{q}, \quad (1 \le i \le e) .$$

Combining the last two inequalities we find a lower bound for $T(x_1,x_2,\ldots,x_{e+1})$. However, from (6.3) and (4.4) we find

$$T(x_1,x_2,\ldots,x_{e+1}) = T\left(\frac{(e+1)!q^{n-e-1}}{|C|}\right) .$$

Combining these results leads to an inequality

$$L(n,q,e) \le R(n,q,e)$$

where, if n is as large as lemma 6.8 insures it should be, $L(n,q,e)$ behaves roughly like $(qn)^{\frac{e-1}{2}}$ and $R(n,q,e)$ behaves like $n(q-1)(e+1)!$ Clearly for $e \ge 4$ this yields a contradiction for sufficiently large n. In fact, a more detailed analysis showed that only a few thousand parameter sets n,e,q were possible. These were handled by computer. The most useful element in the program was the fact that the necessary conditions for the existence of t-designs (in (2.4) λ_i must be an integer) had to be satisfied. Attempts to modify the proof in such a way that $e = 3$ could also be treated were succesful for $q = 2$ only. The extended Golay code

turned out to be the only possibility.
The possibility of the existence of some 3-error-correcting uniformly packed codes with $q > 2$ remains open but it seems doubtful whether they exist.

REFERENCES

1. E. Bannai, On perfect codes in the Hamming schemes H(n,q) with q arbitrary (submitted).
2. L.A. Bassalygo, V.A. Zinoviev, V.K. Leontiev and N.I. Feldman, Nonexistence of perfect codes over some alphabets (in russian), Problemy Peredachi Informatsii 11 (1975), 3-13.
3. D.M. Cvetković and J.H. van Lint, An elementary proof of Lloyd's theorem, Proc. Kon. Ned. Akad. v. Wet. (to appear).
4. P. Delsarte, Two-weight linear codes and strongly regular normed spaces, Discr. Math. 3 (1972), 47-64.
5. ---------, An algebraic Approach to the Association Schemes of Coding Theory, Philips Res. Repts. Suppl. 10 (1973).
6. J.M. Goethals and H.C.A. van Tilborg, Uniformly packed codes, Philips Res. Repts. 30 (1975), 9-36.
7. J.H. van Lint, Recent results on perfect codes and related topics, in Combinatorics I (M. Hall and J.H. van Lint eds.), Math. Centre Tracts 55 (1974).
8. H.F. Reuvers, Some nonexistence theorems for perfect error correcting codes, Thesis, Technological University Eindhoven (1977).
9. N.V. Semakov, V.A. Zinoviev and G.V. Zaitsev, Uniformly packed codes, Problemy Peredachi Informatsii 7 (1971), 38-50.
10. G. Szegö, Orthogonal polynomials, Amer. Math. Soc. Coll. Publ. 23 (1959).
11. A. Tietäväinen, Nonexistence of nontrivial perfect codes in case $q = p_1^r p_2^s$, $e \geq 3$, Discrete Math. (to appear).
12. H.C.A. van Tilborg, Uniformly packed codes, Thesis, Technological University Eindhoven, (1976).

GPSR Compliance

The European Union's (EU) General Product Safety Regulation (GPSR) is a set of rules that requires consumer products to be safe and our obligations to ensure this.

If you have any concerns about our products, you can contact us on ProductSafety@springernature.com

In case Publisher is established outside the EU, the EU authorized representative is:

Springer Nature Customer Service Center GmbH
Europaplatz 3
69115 Heidelberg, Germany

Batch number: 10607222

Printed by Printforce, the Netherlands